PREFACE

It gives me an immense pleasure to introduce my new book entitled **Hybrid Polyaniline Nanocomposite for Humidity Sensing.** The book contains five chapters based on the conducting polymer nanocomposites for humidity sensing application such as Introduction to Nanocomposite, Synthesis of nanocomposites, Characterization techniques, Humidity tests and finally summary of the work. Each chapter has got its own importance in the field of sensor technology, this book will keep the interest of the scientists, academicians and approach of Materials Chemistry and Nanocomposites, main motto of my book is to enhance the knowledge of science for the next generation of the contemporary world. I have made an effort to bring it in the form of systematic approach. The book is useful for the both under graduate and post graduate students as well as for the scientists. The book is very useful to acquire deep knowledge in the field of nanocomposites and humidity sensor.

I am very grateful to Prof. S. R Niranjana Vice- Chancellor. And also am thankful to Registrar Prof. Somashekarappa and Prof. D. M Madari Registrar (Evaluation) for their suggestions and cooperation. I am very thankful to Dr. M.V.N. Ambika Prasad for his support during my PhD research and thereafter. I am thankful to Dr. Anilkumar R Koppalkar and Dr. Ameena Parveen for their support, suggestions and cooperationto bring this book to readers, also thankful to Prof. A Venkataramana, Dean Faculty of Sciences and Professor in Department of Chemistry, Dr. R. L. Raibagkar. Professor in Applied Electronics, Gulbarga University, Kalaburagi for their formal or informal support during the course of research work.

I very thankful to my family members, without their cooperation and support I will become a layman, their nature and attitude supported me to carry out the research in the field of research, so I am very grateful to Parents, brothers and sisters, in laws, and pupils of our family. For their constant help in carrying research activities without their support book in hand would not be possible. I am very thankful to all my students' friends, research scholars of the Department of Materials Science and all friends who have formally or informal helped me during course of my research activities.

Dr. Aashish Roy
Assistant Professor
Department of Industrial Chemistry
Adj. Faculty of Center for Nanotechnology
Addis Ababa Science and Technology
University, Addis Ababa.

FOREWORD

It's a matter of immense significance that Dr. Aashish Roy, Assistant Professor, Department of Industrial Chemistry, Addis Ababa Science and Technology University. who has taken this good effort in the field of research activities. He is young, enthusiastic critical thinker and good academician at national and international. In this book, design andmanufacturing of industrial compounds are well explained. As a young scholar Dr. Aashish Roy is pursuing his academic and research activities in this part of the region, I bless him to success in his academic, research activities and in his carrier.

Dr. Prakash M. Badiger
Historian
Gulbarga University, Kalaburagi-585106.

FOREWORD

It's a matter of immense significance that Dr. Aashish Roy, Assistant Professor, Department of Industrial Chemistry, Addis Ababa Science and Technology University. who has taken this good effort in the field of research activities. He is young, enthusiastic critical thinker and good academician at national and international. In this book, design and manufacturing of industrial compounds are well explained. As a young scholar Dr. Aashish Roy is pursuing his academic and research activities in this part of the region, I bless him to success in his academic, research activities and in his carrier.

Dr. Anilkumar R Koppalkar
Professor
Department of Physics
S.S.Margol College of Arts, Commerce and Science
Gulbarga, Karnataka, India

CONTENTS

CHAPTER I
1.0. Introduction and Definition of Composites

Many of the modern technologies require materials with unusual combinations of properties that cannot be met by the conventional metal, metal alloys, ceramic and polymeric materials. This is especially true for materials that are needed for aerospace, transportation, battery, EMI shielding and sensor applications. For example, electronic engineers and researchers are searching for structural materials that have environmental stability, low density, strong, stiff; abrasion and high impact are not corroded, as well as controllable electrical conductivity and interesting redox properties associated with the chain nitrogen. The electrical properties of the polymers can be improved substantially by secondary doping. Polymer compounds can be designed to achieve the particular conductivity required for a given application.

The development of composites materials is the answer to these. Generally speaking, a composite is considered to be any multiphase material that exhibits significant properties of both constituent phases such that better combinations of properties were realized. According to this principle of combination action, better property combinations are fashioned by the judicious combination of two or more distinct materials.

A composite, in the present context, is a multiphase material that is artificially made, as opposed to one that forms naturally. In addition, the constituent phases must be chemically dissimilar and separated by a distinct interface. Thus, most metallic alloys and many ceramic composites do not fit in this definition because their multiphase formation is a consequence of natural phenomena. Normally most of the composites materials are composed of just two phases; one is termed *as the matrix*,

1

which is continuous and surrounds the other phase, often called as *the reinforcement* as shown in figure 1. The properties of composites are function of the properties of the constituent matrix, their relative amounts and geometry of the dispersed reinforcement. "Dispersed phase geometry" in this context means the shape, size, distribution

and orientation of the particles.

Figure 1 defines the composite

1.1. Classification of Composites: Composit /material is a material composed of two or more distinct phases (matrix phase and dispersed phase) and having bulk properties significantly different from those of any of the constituents.

1.1.1. Matrix phase: The primary phase, having a continuous character, is called matrix. Matrix is usually more ductile and less hard phase. It holds the dispersed phase and shares a load with it.

1.1.2. Reinforcement phase (Dispersed): The second phase is embedded in the matrix in a discontinuous form. This secondary phase is called reinforcement / dispersed phase. This phase is usually stronger than the matrix, therefore it is sometimes called reinforcing phase.

There are two classification systems of composite materials. One of them is based on the material structure and the second is based on the material matrix (metal, ceramic and polymer):

1.2. Classification based on structure

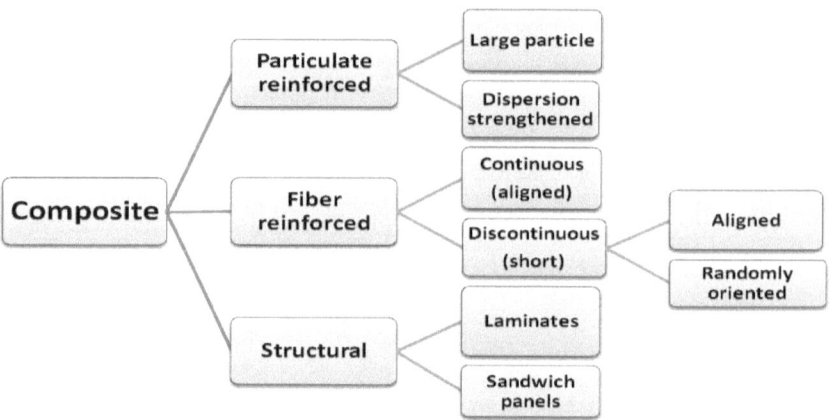

Figure1.1 shows classification scheme for composites based on reinforcement

1.2.1. Particulate Composites

Particulate composites consist of a matrix reinforced by a dispersed phase in the form of particles.

1.2.2. Fibrous Composites: These are classified into two groups

a) Long-fiber reinforced composites (Continuous). Long-fiber reinforced composites consist of a matrix reinforced by a dispersed phase in the form of continuous fibers.

i. Unidirectional orientation of fibers; ii. Bidirectional orientation of fibers (woven).

b) Short-fiber reinforced composites (Discontinuous). Short-fiber reinforced composites consist of a matrix reinforced by a dispersed phase in the form of discontinuous fibers (length < 100nm).Further they are classified into composites with random and preferred orientation of fibers.

1.2.3. Structural Composites

When a fiber reinforced composite consists of several layers with different fiber orientations, it is called multilayer composite.

i.   Laminate   ii. Sandwich

1.3. Classification based on Matrix

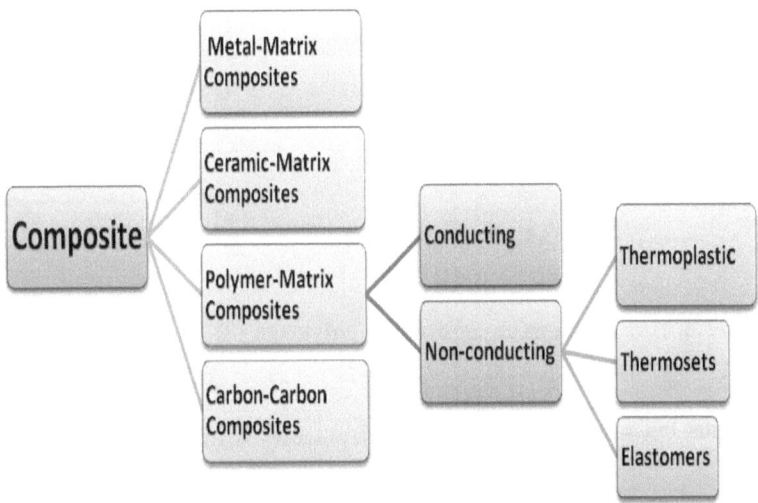

Figure1.2 shows classification scheme for composites based on matrix

1.3.1. Metal Matrix Composites (MMC)

Metal Matrix Composites are composed of a metallic matrix (aluminum, magnesium, iron, cobalt, copper) and a dispersed ceramic (oxides, carbides) or metallic (lead, tungsten, molybdenum) phase.

1.3.2. Ceramic Matrix Composites (CMC)

Ceramic Matrix Composites are composed of a ceramic matrix and embedded fibers of other ceramic material (dispersed phase).

1.3.3. Polymer Matrix Composites (PMC)

Polymer Matrix Composites are composed of a matrix from thermoset (Unsaturated Polyester (UP) Epoxiy (EP)) or thermoplastic (Polycarbonate (PC), Polyvinylchloride, Nylon, Polysterene) and embedded glass, carbon, steel and Kevlar fibers (dispersed phase).

Based on electrical conductivity polymer matrix are further classified as:

i) Non-conducting polymer matrix composites

ii) Conducting polymer matrix composites

1.4. Introduction to conducting polymers

The twentieth century witnessed a tremendous change in the human life style due to the revolutionary discoveries as well as developments in the field of polymer science and engineering. There is a long history of successful development, which came from the enormous contributions of numerous people [1]. The development of conducting polymers became one of the most promising field since the discovery of inherently conducting polymers by Alan J. Heeger, Hideki Shirakawa and Alan G.

MacDiarmid [2] in 1977, for which they were awarded the Nobel Prize in Chemistry, 2000. These polymers are assumed to have tremendous applications in different fields, especially in the electronic industry. But the major concern in this area is the processibility of these polymers as most of them are not stable at processing conditions. Hence, the main objective of the work is the attainment of processibility in these conducting polymers and their blends.

Conducting polymers can be classified mainly into three types:

a) Inherently conducting polymers (ICPs),

b) Conducting polymer composites, and

c) Ionically conducting polymers.

Our field of interest is mainly concentrated on inherently conducting polymers and improvement of its processibility.

1.4.1. Inherently conducting polymers (ICPs)

An organic polymer that possesses the electrical, electronic, magnetic and optical properties of a metal while retaining the mechanical properties, processibility etc. commonly associated with a conventional polymer, is termed an "inherently conducting polymer" (ICP) more commonly known as "synthetic metal"[3] In 1975, the first papers on the novel metallic polymer, poly(sulfur-nitride), $(SN)x$ appeared in the literature. It was in 1976 that A.G.MacDiarmid, H.Shirakawa, A.J.Heeger and coworkers discovered organic conducting polymers and their ability to dope these polymers over the full range from insulator to metal [4]. They have carried out doping experiments on polyacetylene with bromine at the University of Pennsylvania on Tuesday, the 23rd of November 1976 and subsequently with iodine [5]. That event heralded the dawn of a new era of conducting polymers.

The most common examples of inherently conducting polymers are Polyacetylene, Polyaniline, Polypyrrole, Polythiophene, Poly(phenylene), Poly(phenylene vinylene) etc. The Figure 1.3 shows some of the conjugated polymers, which have been studied as intrinsically conducting polymers.

The unique electronic properties of the conjugated polymers are derived from the presence of π - electrons, the wave functions of which are delocalized over a long portion of the polymer chain when the molecular structure of the backbone is planar. It is therefore necessary that there are no large torsion angles at the bonds, which would decrease the delocalization of the π – electrons system [6].

The essential properties of the delocalized π - electrons system, which differentiate a typical conjugated polymer from a conventional polymer with σ - onds are as follows: (a) the electronic (π) band gap (Eg) is relatively small ($\sim$ 1 to 3.5 eV), with corresponding low excitation and semi-conducting behaviour; (b) the polymer molecules can be easily oxidized or reduced, usually through charge transfer with atomic or molecular dopant species, to produce conducting polymers; (c) net charge mobility in the conducting state are large enough so that high electrical conductivities are realized, and (d) quasi – particles, which, under certain conditions, may move relatively freely though the material [7,8].

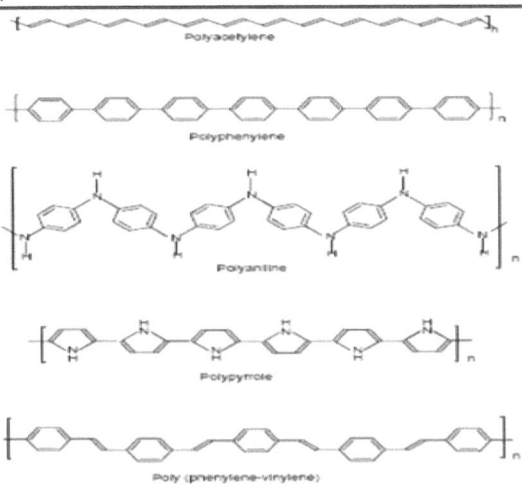

Figure 1.3 shows some of the conjugated polymers

The electrical and optical properties of these materials depend on the electronic structure and basically on the chemical nature of the repeating unit. The general requirement of the electronic structure in these polymers were recognized and described many years ago. The electronic conductivity depends on both the density and drift mobility of the carriers. The carrier drift mobility is defined as the ratio of the drift velocity to the electric field and reflects the ease with which carriers are propagated.

1.4.2. Conducting Polymer Composites:

These are usually prepared by the addition of conducting fillers in the insulating polymer matrix. The conductive fillers commonly used are metal flakes, graphite, conductive carbon black, etc. These fillers are loaded in the common insulating polymers like PP, PVC, LDPE, etc, to get conducting polymer composites or master batches. These are used as semiconducting layers in high voltage applications, EMI shielding materials, etc.

8

The major problem in this area is the processing problems created due to the filler loading. To get sufficient conductivity for these applications, more than20% filler loading is required. This higher addition of these rigid fillers will cause a drastic increment of melt viscosity, which causes serious processing problems. Moreover, these fillers also affect the properties of finished products like aesthetics, brittleness, poor finish, etc. Hence, there is need to develop process where an alternative material or blend can be prepared preferably from conducting polymers.

1.5. Polyaniline

Due to its ease of synthesis and processing, environmental stability, relatively high conductivity and cost economics, polyaniline is probably the most industrially important conducting polymer today [9-11]. Polyaniline is a typical phenylene based polymer having a chemically flexible -NH- group in the polymer chain flanked by phenyl ring on either side. Polyaniline represents a class of macromolecules whose electrical conductivity can be varied from an insulator to a conductor by the redox process. This polymer can achieve its highly conductive state either through the protonation of the imine nitrogens or through the oxidation of amine nitrogens. For example the conducting state of PANI can be obtained in its 50% oxidized emeraldine state in aqueous acids like HCl and the resulting material is a p-type semiconductor [12-14]. With the extent of doping polyaniline can have four different oxidation states as shown in figure 1.4 [15-16] like Leucomeraldine base (LEB), Emeraldine (EB), Emeraldine salt (ES) and Pernigraniline (PNB) .

Figure 1.4 various oxidation states of polyaniline.

Oxidative doping of the leucomeraldine base or protonic acid doping of the emeraldine base material produces the conducting emeraldine salt whose conductivity varies between 0.5 S/cm and 400 S/cm depending on the methods of preparation. Extensive studies of the emeraldine salt (ES) material have shown that the metallic state is governed by inhomogeneous disorder. In the conducting state, there are regions that are three-dimensionally ordered in which the conducting electrons are three-dimensionally delocalized and regions where the polymer is strongly disordered, in which conduction electrons diffuse through one-dimensional polymer chains that are nearly electrochemically isolated. One dimensional localization in these nearly isolated chains lead to decrease in conductivity with decrease in temperature.

Polyaniline can be synthesized mainly by chemical or electrochemical oxidation of aniline under acidic conditions. The method of synthesis depends on the intended application of the polymer. For bulk production chemical method, where as for thin films electrochemical method is preferred.

1.5.1. Chemical Synthesis:

The conventional method of synthesis of emeraldine salt is the emulsion polymerization of aniline monomer in aqueous media in presence of a mineral acid like HCl [17-20]. An oxidant like ammonium per sulphate or potassium dichromate can be used to initiate the reaction [21- 23]. The ideal molar ratio of monomer to acid to oxidizing agent is proved to be 1: 1: 1 [24, 25]. The aniline salt of protonic acid in the protonic acid medium is mixed with aqueous solution of ammonium per sulphate with a continuous stirring for 4 hrs. The precipitate obtained is then filtered and washed with distilled water so as to obtain emeraldine salt.

The principle function of the oxidant is to withdraw a proton from the aniline monomer. The polymerization reaction is summarized as follows:

$$Ar-NH_2 \xrightarrow{ArNH_2} Ar-NH-Ar-NH_2 \xrightarrow[2e, -2H^+]{Ar-NH_2}$$

$$Ar-NH-Ar-NH-Ar-NH_2 \xrightarrow[\text{coupling}]{\text{Further oxidative}} Polyaniline$$

The factors affecting the polymerization process are the pH of the solution, type of the acids used, its concentration, effective size, solvation and electronegativity of the conjugate base associated with a given acid. [26].

1.5.2. Polymerization Mechanism

The wide variety of methods employed for preparation of PANI leads to products whose nature and properties differ greatly. The mechanism and kinetics of PANI formation has been extensively studied for the identification of the intermediates and the steps involved. This knowledge is essential to correlate the relationships between possible reaction pathway and properties of the polymeric products [27]. Information concerning the mechanism of PANI formation has invariably been gathered with the aid of electro-chemical

methods. Various polymerization mechanisms and electrochemical aspects of

the formation of PANI have been proposed by different authors depending on

the protocol used in the synthesis of PANI and have been reviewed in detail

[28-30].The polymerization reaction is a self-catalyzing reaction and obeys the

law: $i/nFA = K_C$, where K_C is the autocatalytic rate constant and has a value of

~0.47 s^{-1} for a 140 nm thick PANI film [31]. The most accepted first step in the

reaction mechanism is the formation of radical cations which is resonance

stabilized by several canonical forms and is represented in figure 1.4.

Figure 1.4 shows the reaction mechanism of polymerization

1.5.3. Dopants for Polyaniline:

Polyaniline can be easily doped by non-redox doping method. Polyaniline

holds a special position amongst conducting polymers in that it's highly conductive

doped form is accessible by two completely different processes – protonic acid doping

(non-redox) and oxidative doping. Protonic acid doping of emeraldine base units,

results in the complete protonation of the imine nitrogen atoms gives fully protonated

emeraldine salt.

Doped polyaniline can be obtained by chemical oxidation (p-doping) of

leucomeraldine base. This actually involves the oxidation of the σ/π-system rather

than just the π-system of the polymer, as is usually the case of p- doping.

There are plenty of reports available on different acid as well as ester dopants used for polyaniline. They include common mineral acids, high molecular weight long-chain organic sulfonic acids, phosphoric acids and their esters, etc. The standard route of synthesis of doped polyaniline is the one with Hydrochloric acid (HCl) as dopant ions as reported by many authors. The other mineral acids commonly used include Sulfuric acid (H_2SO_4), Hydrofluoric acid (HF) Perchloric acid ($HClO_4$) etc. The low molecular weight organic acids include Formic acid, Acetic acid, Acrylic acid, etc. High molecular weight long-chain organic sulfonic acids include Camphor sulfonic acid (CSA) [32-36] Methane sulfonic acid (MeSA) [37-39] p-Toluene sulfonic acid (PTSA), [40-42] dodecylbenzenesulfonic acid (DBSA) [43-52] polystyrenesulfonic acid (PSSA) [53-55], etc. Other than the above-mentioned acid dopants, esters of phosphoric acid [56] as well as phthalic acid [57] were also used as effective non-redox dopant ions. Some examples of the organic sulfonic acid dopants used are shown in the figure 1.5.

Camphor Sulfonic Acid (CSA)   Dodecyl Benzene Sulfonic Acid (DBSA)

Polystyrene Sulfonic Acid (PSSA)

Figure 1.5 represents the structures of the sulphonic acid are used as dopant ions for polyaniline.

It was seen that the dopant ion could alter the properties of the conducting polyaniline formed. It can affect properties like conductivity, crystallinity, solubility and, hence, processibility. Amongst these, the organic sulfonic acids have been found to impart many interesting properties like high solubility and thermal stability to polyaniline. The sulfonic acids like DBSA, PSSA, etc., which have long-chain alkyl groups can plasticize the polyaniline formed by increasing the interchain separation, thereby facilitating the solvent molecules penetrating into the polyaniline lattice. Also, the alkyl chain can act as an effective solvent compatible group to improve the solubility of the polymer. These groups can also acts as compatibilizers as well as plasticizers when blended with other polymers. In the present work hydrochloric acid was selected as dopant for polyaniline. Apart from this, the author employs zinc ferrite, nickel zinc oxide and copper oxide as secondary dopants. The efficiency of these dopants to provide better properties and processibility to polyaniline were examined.

1.5.4. Mechanism of Conductivity

The electrical conductivity of doped conducting polymers can be varied up to metallic state in the range of more than ten orders of magnitude. Charge transport in these polymers has been extensively investigated but still remains poorly understood [58]. The conjugation arising from chemical unsaturation of the carbon atom in conducting polymers is the main cause for charge transport. The presence of localized electronic states of energies less than the band gap arising due to the changes in the local bond order have lead to the possibility of new types of charge conduction phenomenon in these conjugated polymers. All conducting polymers have intrinsic topological defects which are introduced during polymerization and their ground states are non-degenerate. Removal of a charge from the valance band generates radical cations whose energy lies in the band gap. In solid state physics, such a radical cations which is partially delocalized over some polymer segments is called a 'polaron'. It stabilizes itself by polarizing the medium around it and hence its name. Formation of polaron is associated with the distortion of lattice and presence of two localized electronic states in the gap. According to the model proposed by Brazovski - Kirova, formation of a polaron leads to the possibility of three new optical transitions (figure 1.6).

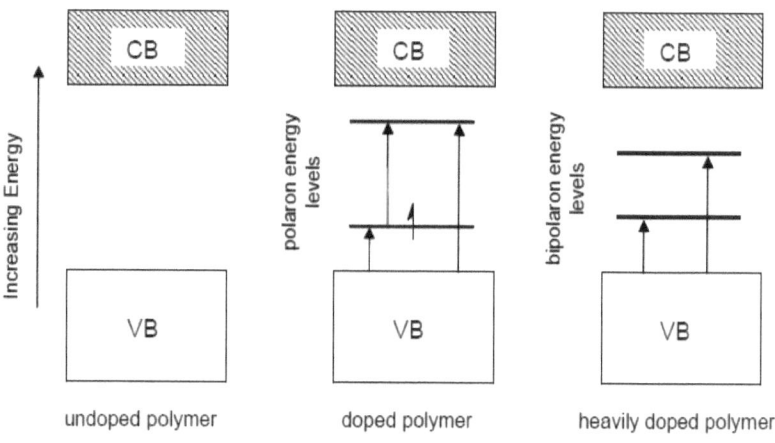

VB = valence band; CB = conduction band

Figure 1.6 Illustration of energy levels and allowed transitions of polarons and bipolarons.

When a second electron is removed from the system, it may come from either a different segment of the polymer chain creating another polaron, or from the first polaron to generate a direction which, in solid state physics, is referred to as a bipolaron (figure 1.4). A bipolaron is also associated with structural deformation and two charges are not independent but act as a pair.

15

Application of an external electric field makes both polaron and bipolaron mobility via the rearrangement of conjugation.

In the case of PANI, the charged species are formed during protonation of the polymer, which are subsequently responsible for the increase in conductivity [59]. Several other mechanisms are also proposed for the conductivity of PANI. Salaneck and coworkers [60,61], based on temperature dependent conductivity studies, have proposed a one dimensional variable range hopping or three-dimensional fluctuation induced tunneling models.

1.5.5. Applications of processible Conducting Polymers:

Conducting polymers are assumed to have tremendous applications in various fields especially in electronic appliances. They have importance in large number of application like LED display, conductive surfaces, solid phase sensors, optical storage lithographs, transducers, EMI/ESD shielding, plastic batteries, membrane for gas separation, magnetic recording, etc. But as discussed the problem hindering these wonderful applications is the poor processibility in these polymers. Improvement of the processibility will enable scientists and technologists to explore this to create a new looking world of conducting polymers. Some of the possible applications of processibile conducting polymers are discussed below.

By coating an insulator with a very thin layer of conducting polymer it is possible to prevent the buildup of static electricity. This is particularly important where such a discharge is undesirable. Such a discharge can be dangerous in an environment with flammable gasses and liquids and also in the explosives industry. In the computer industry the sudden discharge of static electricity can damage microcircuits. This has become particularly acute in recent years with the development of modern integrated circuits. To increase speed and reduce power consumption, junctions and connecting lines are finer and closer together. The resulting integrated circuits are more sensitive and can be easily damaged by static discharge at a very low voltage. By modifying the thermoplastic used by adding a conducting plastic into the resin results in a plastic that can be used for the protection against electrostatic discharge [62]. The solution processible polyaniline blends can contribute to this application.

Many electrical devices, particularly computers, generate electromagnetic radiation, often radio and microwave frequencies. This can cause malfunctions in nearby electrical devices. The plastic casings used in many of these devices are transparent to such radiation. By coating the inside of the plastic casing with a conductive surface this radiation can be absorbed. This can best be achieved by using conducting plastics. This is cheap, easy to apply and can be used with a wide range of resins. The final finish generally has good adhesion, gives a good coverage and thermally expands approximately the same as the polymer it is coating, needs just one step and gives a good thickness. Processible conductive polyaniline and blends either solution processible or, more promisingly the melt processible blends can provide wonderful material for EMI shielding.

Corrosion is a part of our everyday life. When metal donates some of its electrons to oxygen it causes the formation of impurities that weaken the structure, which is known as corrosion. Painting a metal with zinc or covering the metal with plates of zinc can delay corrosion. Zinc, being more reactive and a good electron donor reacts with oxygen more readily and the metal underneath is not affected. Unfortunately, coatings and plates of zinc cannot last long. All of the problems caused by corrosion can be reduced considerably by a wonderful "plastic coating that virtually eliminates rust and corrosion-which could help cars, bridges, pipelines and other metal structures last up to 10 times longer" This coating can be prepared by using polyaniline. Coatings like zinc paint create a physical barrier, but polyaniline works completely different. Polyaniline is a "catalyst that mediates the reaction that leads to rust". Polyaniline halts corrosion by "accepting electrons from the metal and in turn, donates them to oxygen" creating a two-step reaction that "forms a layer of pure iron oxide". In the laboratory, under controlled conditions "polyaniline prevented rust 10,000 times more effectively than zinc" and in the field polyaniline "proved three to 10 times more effective" reported Bernard Wessling president and managing partner of Ormecon Chemie GmdH & Co in Ammersbek, Germany. The polymer coating polyaniline is not a heavy metal and does not propose a threat to human health and "it is also cheaper than zinc", Wessling described it as an "organic metal" that could last forever. Processible polyaniline or blends can provide a promising material in this respect.

In high voltage cables, there is a high probability of occurrence of spark due to the accumulation of heavy static charges on the insulating material used as sheathing for the cable. It can lead to hazardous fire and cause damages for cable system and neighboring materials. To avoid this, there used to be a semiconducting inner layer for the sheathing in contact with the actual conductor. This gives a gradual reduction of conductivity fro the conductor to the outer insulating sheath, thereby, facilitating the discharge of the static charge accumulation and protects from any sparking problems. This semiconducting layer is usually made by conducting carbon black composites in polymers like LDPE and PVC. It needs 15 - 20% of carbon black loading to obtain useful conductivity, which leads to considerable processibility problems due to the dramatic increase in melt viscosity while extrusion. If carbon black could be replaced by conducting polymers like polyaniline, it can reduce the above discussed processing problem. For this, the conducting polymer has to be thermally stable (retaining of conductivity) at processing temperatures and, hence, melt processible. So the melt processible conducting polymer or its blends offer great potential in high voltage cables applications.

1.6. Introduction to Ferrites:

Ferrites are chemical compounds consisting of ceramic materials with iron (III) oxide (Fe_2O_4) as their principal component [63]. Many of them are magnetic materials and they are used to make permanent magnets, ferrite cores for transformers, and in various other applications.

Many ferrites are spinels with the formula AB_2O_4, where A and B represent various metal cations, usually including iron. Spinel ferrites usually adopt a crystal motif consisting of cubic close-packed (fcc) oxides (O^{2-}) with A cations occupying one eighth of the tetrahedral holes and B cations occupying half of the octahedral holes—that is, the inverse spinel structure.

The magnetic material known as "ZnFe" has the formula $ZnFe_2O_4$, with Fe^{3+} occupying the octahedral sites and half of the tetrahedral sites. The remaining tetrahedral sites in this spinel are occupied by Zn^{2+}[64].

Some ferrites have hexagonal crystal structure, e.g. barium ferrite $BaO:6Fe_2O_3$ or $BaFe_{12}O_{19}$. Ferrites are usually non-conductive ferrimagnetic ceramic compounds derived from iron oxides such as hematite (Fe_2O_3) or magnetite (Fe_3O_4) as well

19

as oxides of other metals. Ferrites are, like most other ceramics, hard and brittle. In terms of their magnetic properties, the different ferrites are often classified as "soft" or "hard", which refers to their low or high magnetic coercivity.

1.6.1. Soft ferrites:

Ferrites that are used in transformer or electromagnetic cores contain nickel, zinc, and/or manganese compounds. They have a low coresivity and are called soft ferrites.

The low coresivity means the material's magnetization can easily reverses its direction without dissipating much energy (hysteresis losses), and the material's high resistivity prevents eddy currents in the core, thus it is an another source of energy loss. Because of their comparatively low losses at high frequencies, they are extensively used in the cores of RF transformers and inductors in applications such as switching-mode power supplies.

The most common soft ferrites are: Manganese-zincferrite ($Mn_aZn_{(1-a)}Fe_2O_4$). MnZn have higher permeability and saturation induction than NiZn. Nickel-zinc ferrite ($Ni_aZn_{(1-a)}Fe_2O_4$). NiZn ferrites exhibit higher resistivity than MnZn, and therefore more suitable for frequencies above 1 MHz.

1.6.2. Hard ferrites:

In contrast, permanent ferrite magnets are made of hard ferrites, which have a high coercivity and high retentivity. These are composed of iron and barium or strontium oxides. The high coercivity means the materials are very resistant to becoming demagnetized, an essential characteristic for a permanent magnet. They also conduct magnetic flux well and have a high magnetic permeability. This enables these ceramic magnets to store stronger magnetic fields than iron itself. They are cheap, and are widely used in household products such as refrigerator magnets. The maximum magnetic field B is about 0.35 tesla and the magnetic field strength H is about 30 to 160 kilo-ampere turns per meter (400 to 2000 Orested) [65]. The density of ferrite magnets is about $5g/cm^3$.

The most common hard ferrites are; Strontium ferrite, $SrFe_{12}O_{19}$ ($SrO6Fe_2O_3$), a common material for permanent magnet applications.

Barium ferrite, $BaFe_{12}O_{19}$ ($BaO\ 6Fe_2O_3$), a common material for permanent magnet applications. Barium ferrites are robust ceramics that are generally stable to moisture and corrosion-resistant. They are used in e.g. subwoofer magnets and as a medium for magnetic recording, e.g. on magnetic stripe cards.

Cobalt ferrite, $CoFe_2O_4$ ($CoO\ Fe_2O_3$), used in some media for magnetic recording [66].

1.6.3. Uses of Ferrites

- Ferrite cores are used in electronic inductors, transformers and electromagnets where the high electrical resistance of the ferrite leads to very low eddy current losses. They are commonly seen as a lump in a computer cable called a ferrite bead, which helps to prevent high frequency electrical noise (radio frequency interference) from exiting or entering the equipment.

- Early computer memories stored data in the residual magnetic fields of hard ferrite cores, which were assembled into arrays of core memory. Ferrite powders are used in the coatings of magnetic recording tapes. One such type of material is iron (III) oxide.

- Ferrite particles are also used as a component of radar-absorbing materials or coatings used in stealth aircraft and in the absorption tiles lining the rooms used for electromagnetic compatibility measurements.

- Most common radio magnets, including those used in loudspeakers, are ferrite magnets. Ferrite magnets have largely displaced Alnicomagnets in these applications.

- It is a common magnetic material for electromagnetic instrument pickups, because of price and relatively high output. However, such pickups lack certain sonic qualities found in other pickups, such as those that use Alnico alloys or more sophisticated magnets.

- Ferrite nanoparticles exhibit superparamagnetic properties.

1.6.4. Zinc Ferrite

Zinc ferrites are a series of synthetic inorganic compounds of zinc and iron (ferrite) with the general formula of $Zn_xFe_{3-x}O_4$. Zinc ferrite compounds can be prepared by aging solutions of $Zn(NO_3)_2$, $Fe(NO_3)_3$, and triethanolamine in the presence and in the absence of hydrazine, or reacting iron oxides and zinc oxide at high temperature. Spinel (Zn, Fe) Fe_2O_4 appears as a tan-colored solid that is insoluble in water, acids, or diluted alkali. Because of their high opacity, zinc ferrites can be used as pigments, especially in applications requiring heat stability. For example, zinc ferrite prepared from yellow iron oxide can be used as a substitute for applications in temperatures above 177 °C. When added to high corrosion-resistant coatings, the corrosion protection increases with an increase in the concentration of zinc ferrite. A recent investigation shows that the zinc ferrite, which is paramagnetic in the bulk form, becomes ferrimagnetic in nanocrystalline thin film format. A large room temperature magnetization and narrow ferromagnetic resonance line width have been achieved by controlling thin films growth conditions.

1.7. Introduction to metal oxide

Metal oxides constitute the most fascinating class of materials, exhibiting a variety of structures and properties [67]. The metal oxygen bond can vary anywhere between highly ionic to covalent or metallic. The unusual properties of transition metal oxides are clearly due to the unique nature of the outer d-electrons. The phenomenal range of electronic and magnetic properties, exhibited by transition metal oxides is noteworthy. Thus, the electrical resistivity in oxide materials spans the wide range of 10^{-10} to 10^{20} Ω cm. We have oxides with metallic properties (e.g. RuO_2 RuO_3) at on end of the range and oxides with highly insulating behavior (e.g. $BaTiO_3$) at the other as shown in fig. 1.9.2 (a) & (b)

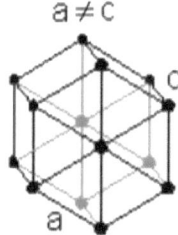

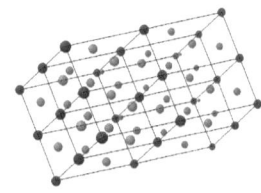

Figure 1.9.2(a) Rutile structure   **1.9.2(b) Perovskite   structure**

(highly metallic)   **(highly insulating)**

There are also oxides that transverse either these regimes with changes in temperature, pressure or composition (e.g. V_2O_5, $La_{1-x}Sr_xVO_3$). Interesting electronic properties also arise from charge density wave (e.g. $K_{0.3}MoO_3$), charge ordering (e.g. Fe_3O_4) and defect ordering (e.g. $Ca_2Mn_2O_5$, $Ca_2Fe_2O_5$). Oxides with diverse magnetic properties anywhere from ferromagnetism (e.g. CrO_2, $La_{0.5}Sr_{0.5}MnO_3$) to anti-ferromagnetism (e.g. NiO, $LaCrO_3$, α-Fe_2O_3) are known. Many oxides posses switchable orientation states as in ferroelectric (e.g. $BaTiO_3$, $KNbO_3$) and ferroelastic [e.g. $Gd_2(MoO_4)_3$] materials. Then, there is a variety of oxides bronzes showing a gamut of property [68].

The unusual properties of transition metal oxides that distinguish them form different phases are due to several factors:

1. Oxides of d-block transition elements have narrow electronic bands, because of the small overlap between the metal d-orbital and the oxygen p-orbital. The bandwidths are typically of the order of 1-2 eV (rather the 5-15 eV as in most metals).

2. Electron correlation effects play an important role, as expected because of the narrow electronic bands. The local electronic structure can be described in terms of atomic like states [e.g. Cu^+ (d^{10}), Cu^{2+} (d^9) and Cu^{3+} (d^8) for Cu in CuO] as in the Heitler-London limit.

3. The polarizability of oxygen is also of importance. The divalent oxide ion O^{2-} does not exactly describe the state of oxygen and configurations such as O^- have to be included especially in the solid state which gives rise to polaronic and bipolaronic effects. Species, such as O^- which are oxygen holes with a p^5

configuration instead of filled p^6 configuration of O^{2-}, can be made mobile and correlated.

4. Many transition metal oxides are not truly three-dimensional but also have low-dimensional features [69]

Among the transition metals oxides, zinc oxide (ZnO) [70], aluminum oxide (Al_2O_3), titanium oxide (TiO_2) [71], tin oxide (SnO_2) [72], tungsten oxide (WO_3), Vanadium oxide (V_2O_5), cerium oxide (CeO_2), iron oxide (Fe_2O_3), cobalt oxide (Co_3O_4) [73] etc. are mostly widely known oxides and industrially employed transition metal oxides since last fifty years. The cause of these oxides to have become important both scientifically and industrially include their applications for sound and picture recording, data storage, humidity and gas sensors, conducting composite super capacitors, electrochromic display devices, etc. In the present study the following transition metal oxides are used.

1|Nickel doped Zinc oxide

2|Copper oxide

1.7.1. Nickel doped Zinc oxide:

Metal oxides have attracted a wide attention due to their unique properties and immense potential application in device fabrication [74-77]. Zinc oxide (ZnO) has a direct wide band gap (3.4 eV at Room temperature), which is n-type semiconductor. In ambient condition, ZnO has a stable hexagonal wurtzite structure with lattice spacing $a = 0.325$ nm and $c = 0.521$ nm and composed of a number of alternating planes with tetrahedrally-coordinated O^{2-} and Zn^{2+} ions, stacked alternately along the c-axis. It has attracted intensive research effort for its unique properties and versatile applications in transparent electronics, ultraviolet (UV) light emitters, piezoelectric devices, chemical sensors and heterogenous photocalysts [78-88]. It has been proposed to be a more promising UV emitting phosphor than GaN because of its larger exciton binding energy (60 meV). All these predominant properties make ZnO a great potential in the field of nanotechnology.

Various chemical methods have been developed to prepare nanoparticles of different materials of interest. Most of the ZnO crystals have been synthesized by traditional high temperature solid state method in which, it is difficult to control the particle properties and also energy consuming. ZnO nanoparticles can be prepared on a large scale at low cost by simple solution based method, such as chemical precipitation, sol-gel synthesis, and hydrothermal reaction [89-95]. Many of the earliest synthesis of nanoparticles were achieved by coprecipitation of sparingly soluble products from aqueous solution followed by thermal decomposition of those products to oxides. Coprecipitation method is a promising alternative synthetic method because of the low process temperature and easy to control the particle size. Some of the most commonly substances used in coprecipitation operations are hydroxides, carbonates, sulphates and oxalates.

In the present work Ni doped ZnO particles were synthesized by using Sol-Gel method, which is robust and reliable to control the shape and size of particles without requiring the expensive and complex equipments.

1.7.2. Copper oxide

Copper (II) oxide or cupric oxide (CuO) is the higher oxide of copper. As a mineral, it is known as tenorite.

Copper (II) oxide belongs to the monoclinic crystal system, with a crystallographic point group of 2/m or C_{2h}. The space group of its unit cell is C2/c, and its lattice parameters are a = 4.6837(5), b = 3.4226(5), c = 5.1288(6), $\alpha = 90°$, $\beta = 99.54(1)°$, $\gamma = 90°$. The copper atom is coordinated by 4 oxygen atoms in an approximately square planar configuration [96].

Cupric oxide is used as a pigment in ceramics to produce blue, red, and green (and sometimes gray, pink, or black) glazes. It is also used to produce cuprammonium hydroxide solutions, used to make rayon. It is also occasionally used as a dietary supplement in animals, against copper deficiency. Copper(II) oxide has application as a p-type semiconductor, because it has a narrow band gap of 1.2 eV. It is an abrasive used to polish optical equipment. Cupric oxide can be used to produce dry cell batteries. It has also been used in wet cell batteries as the cathode, with lithium as an anode, and dioxalane mixed with lithium perchlorate as the electrolyte. Copper(II) oxide can be used to produce other copper salts. It is also used when welding with copper alloys.

1.7. Literature Review

Jing Wang et.al [97] has carried out systematic study of Lanthanum ferrite/polymer quaternary acrylic resin for humidity sensing by citrate method. They investigated the electrical property of this humidity sensor, including the resistance versus RH, humidity hysteresis, response – recover time and long term stability. McGovern et.al. [98] and used as a sensing medium in the construction of a resistance-based humidity sensor. The sensors had an overall final thickness of less than 150μm and showed high sensitivity, low resistance, and good reversibility without hysteresis. Despite good progress in the study of charge transfer in conducting polymers, factors affecting electrical conductivity in terms of device applications are not entirely understood. Recently the electrical and humidity sensing properties of polyaniline/WO_3 composites and Humidity Sensing and Electrical Properties of Polyaniline/Cobalt Oxide Composites was studied by Narsimha Parvatikar et.al [99,100].

Labidi. et.al., showed WO_3 based sensors and could be used to detect ethanol. This work proposes to couple dc measurements with ac spectroscopy in order to determine the regions of the device (electrodes, grains or grain boundaries), which are crucial for the detection of ethanol [101]. Wei et al., [102] also revealed that doping of even a very small amount of single-wall carbon nanotubes (SWNTs) could effectively improve the room temperature sensitivity of SnO_2 sensors towards NO_2. Tetsuya Kida et.al., developed the zirconia-based sensing device, fitted with a porous MoO_3 electrode prepared by calcination of a molybdenum polyoxometallate–hexylamine hybrid film, showed considerably high sensitivity toward H_2 (50–500 ppm) and NH_3 (10–100 ppm) in air at 500 °C. It was found that loading Pt onto the MoO_3 electrode improved the recovery behaviors of the sensor toward NH_3, but decreased the sensitivity due to the catalytic combustion of NH_3 over Pt [103].

S. Barazzouk et.al., studied A sensitive MoO_3-based sensor for NO, NO_2 and CH_4 gases has been fabricated. The change in the electrical resistance of the material serves as a sensing parameter. In case of NO and NO_2 gases, the resistance of the sensor element increased and attained a saturation value rapidly. The resistance regained its original value when the gases were turned off. The present sensor has also been found to be suitable for the detection of methane gas. The transient response of the sensor to methane gas, however, is a mirror image of that for NO, NO_2 gases. The reducing gas (CH_4) helps liberate free electrons and decreases the resistivity of the sensor whereas the oxidizing gases (NO_x) take up free electrons leading to an increase in the resistivity of the sensor [104]. Abhilasha Srivatsav et.al., have studied the synthesis of nanostructured SnO_2 by microwave technique and the gas sensitivity measurements showed that, the new aqueous route is better in terms of sensitivity as well as response and recovery time [105].

Q.Y. Soundararajah. et al., have developed the interest in clay polymer nanocomposites (CPN) materials, and have attracted great interest, both in industry and in academia, because they often exhibit remarkable improvement in materials' properties when compared with virgin polymer or conventional micro- and macro-composites. These improvements can include high moduli, increased strength and heat resistance, decreased gas permeability and flammability, optical transparency and increased biodegradability of biodegradable polymers. Such enhancement in the properties of nanocomposites occurs mostly due to their unique phase morphology and improved interfacial properties [106].

Peng Liu has synthesized a series of polyaniline/silica nanosheet composites (PANI/SNS) with different contents of the silica nanosheets derived from vermiculite via acid-leaching were prepared via the insitu chemical oxidation polymerization. It is interesting to note that the electrical conductivities of the PANI/SNS composites increased with the increasing of the contents of the silica nanosheets added because of the moisture absorption [107]. A. L. Kukla et.al., studied the use of thin conducting polymer film (Polyaniline and poly-3-methylthiophene treated with monovalent anions as a sensitive layer in the sensor arrays for chemical recognition of volatile organic analyte is investigated [108]

1.8. Objectives and scope of the work:

Attainment of real processibility for inherently conducting polymers was the main aim of the present work. Although there were some investigations in the past, there was hardly any success in this regard. Till now, the melt processing of these polymers or their blends has not found real success mainly due to the loss of conductivity of these polymers at the processing temperatures. The present studies are directed towards improving the processibility and stability of the conducting polyaniline by suitable modifications such as blending and using dopants that act as compatibilizers or plasticizers. Such modifications can lead to changes in crystalline structure, morphology, electrical properties and optical properties. Hence detailed studies are needed in these aspects as well.

In order to improve the processibility and thermal stability of conducting polyaniline, three different methods were adopted for modifying the polymer: (a) Doping the polymer with molecules having electron acceptors and long alkyl chains, (b) Solution blending of modified polyaniline with thermoplastics having good solubility and processibility in common organic solvents such as PVC, PMMA, PC etc. and, (c) Melt blending of polyaniline with polymers with low processing temperature such as LDPE.

The thermal stability, solubility, compatibility as well as conductivity and, hence the processibility of polyaniline could be improved tremendously by incorporation of a long-chain organic acid as the dopant ion. It was also expected that the single step, direct doping procedure could be used for synthesizing these polyanilines with different long-chain dopant ions instead of the standard route of synthesizing, to avoid tedious steps involved like de-doping, re-doping etc. The conductivity as well as the solubility of polyaniline can also be improved by using these long chain acid dopants. So PANI doped with one of this long chain acid dopant was selected as the ideal candidate for further studies on processibility.

This thermally stable polyaniline was melt-blended with polymers like LDPE and its real processibility was checked in terms of ease of processing and thermal stability of conductivity. The effects of polymeric dispersing agents like PEO and PEG that are ionically conductive in nature were explored as processing aids, with an expected improvement of processibility along with electrical properties of these blends. This blends were identified various applications such as a semiconducting layer in high voltage cables and various ESD / EMI shielding applications.

Solubility of the polyaniline can also be improved by the use of long chain acid dopants. This polyaniline with improved solubility in common organic solvents can be solution blended with soluble polymers like PMMA, PVC, PC, etc. These blends could also provide processible conducting material with better optical properties. These blends could also be used for anti-static coatings, corrosion protective coatings, coating for EMI shielding, etc.

REFERENCES:

1. A.J. Heeger, (Nobel Lecture) Angew. Chem. Int. Ed., 40, (2001), 2591.

2. H. Shirakawa, E.J. Louis, A.G. MacDiarmid, C.K. Chiang and A.J. Heeger, Chem. Commun., (1977) 578.

3. A.G. MacDiarmid, (Nobel Lecture), Angew. Chem. Int. Ed., 40, (2001) 2581.

4. C.K. Chiang, C.R. Fincher Jr., Y.W. Park, A.J. Heeger, H. Shirakawa and E.J.Louis, Phys. Rev. Lett., 39, (1977) 1098.

5. H. Shirakawa, (Nobel Lecture), Angew. Chem. Int. Ed., 40, (2001) 2574.

6. J.L. Bredas and R. Silbey, Conjugated polymers, (Kluwer Dodrecht, 1991)

7. A.J. Heeger, S. Kivelson, Schrieffer and W.P. Su, Rev. Mod. Phys., 60, (1988) 781.

8. W.R. Salaneck and J.L. Bredas, Solid State Comm., 92(1994)31.

9. J.Y. Shimano and A.G. MacDiarmid, Synth. Met., 123(2001)251.

10. E.M. Genies, A. Boyl, M. Lapkowski and C. Trintavis, Synth. Met., 36, (1990)139.

11. Y. Cao, A. Andretta, A.J. Heeger, P. Smith and Polymer, 30(1989)2305.

12. J.-C. Chiang and A.G. MacDiarmid, Synth. Met., 13(1986) 93.

13. J.L. Cadenas and H. Hu, Solar Energy Mate. Sol. Cells, 55(1998)105.

14. A.G. MacDiarmid, J.-C. Chiang and A.F.Richter, Synth. Met., 18, (1987), 317.

15. A.G. Green and A.E. Woodhead, J. Chem. Soc, (1910)1117.

16. A.G. Green and A.E. Woodhead, J. Chem. Soc, (1910)2388.

17. A. G. Green and A. E. Woodhead, J. Chem. Soc., 101(1912)1117.

18. A. G. MacDiarmid and A. J. Epstein. Faraday Discuss. Chem. Soc. 88 (1989)317.

19. J. Stejskal, P. Kratochvil, A. D. Jenkins. Polymer 37 (1996) 367.

20. D. C. Trivedi. Handbook of Organic Conductive Molecules and Polymers, H. S. Nalwa (Ed) Wiley, Chichester, 2 (1997) 505.

21. N. Gospodinova and L. Terlemezyan. Prog. Polym. Sci. 23 (1998) 1443.

22. B. G. Levi. Phys. Today 53 (12) (2000) 19.

23. A. G. MacDiarmid. Angew. Chem., Int. Ed. 40, (2001) 2581.

24. Z. Jin, Y. Su, Y. Duan. Sensor & Actuators B 72 (2001) 75.

25. P. T. Sotomayor, I. M. Raimundo, Jr., A. J. G. Zarbin, J. J. R. Rohwedder, G. O. Netto, O. L. Alves. Sensor & Actuators B 74 (2001) 157.

26. L. A. P. Kane-Maguire and G. G. Wallace. Synth. Met., 39 (2001) 119.

27. R. Holze, In Adv. Funct. Mol. & Poly, H. S. Nalwa (ed.), Gordon and Breach, Tokyo, 2 (2001) 171.

28. A. G. Green and A. E. Wood head, J. Chem. Soc., 101(1912)1117.

29. J. C Chiang, A. G. MacDiarmid, Synth. Met., 13(1986)193.

30. E. M. Genies, A. Boyle, M. Lapkowski and C. Tsintavis, Synth. Met., 36 (1990)139.

31. Y. S. Negi and P. V. Adhyapak, J. Macromol. Sci. - Polym. Rev., C42, 2002, 35.

32. H. Karami, M.F. Mousavi and M. Shamsipur, J. Power Sources, 124, (2003), 303.

33. Y.M. Lee, S.Y. Nam and S.Y. Ha, J. Memb. Sci., 159, (1999), 41.

34. V.V. Chabukswar, S. Pethkar and A.A. Athwale, Sensors and Actuators B: Chem., 77, (2001), 657.

35. A.A. Athwale, M.V. Kulkarni and V.V. Chabukswar, Mate. Chem. Phys., 73 (2002) 106.

36. N.S. Saricifti, L. Smilowitz, Y. Cao and A.J. Heeger, Synth. Met., 55, (1993), 188.

37. Y.Z. Wang, J. Joo, C.-H. Hsu and A.J. Epstein, Synth. Met., 68(1995)207.

38. J.R. Santos Jr., A.J. Motheo, J.A. Malmonge, Y.P. Mascarenhas, L.H.C. Mattoso and A.J.G.C. Silva, Synth. Met., 69, (1995), 141.

39. E.R. Holland, S.J. Pomfret, P.N. Adams, L. Abell and A.P. Monkman, Synth. Met., 84, (1997), 777.

40. Y.F. Nicolau, P.M. Beadle and E. Banka, Synth. Met., 84, (1997), 585.

41. D. Berner, M. Nechtschein, P. Rannou, J.-P. Travers, J. Davenas and D. Djurado, Synth. Met., 101, (1999), 727.

42. M. Reghu, Y. Cao, D. Moses and A.J. Heeger, Synth. Met., 57(1993),5020.

43. V.G. Kulkarni, L.D. Campbell and W.R. Mathew, Synth. Met.,30 (1989), 321.

44. S.K. Jeong, J.S. Suh, E.J. Oh, Y.W. Park, C.Y Kim and A.G. MacDiarmid, Synth .Met., 69, (1995), 171.

45. T.-C. Wen, J.-B. Chen and A. Gopalan, Met. Lett., 57(2002)280.

46. W.A.Jr Gazotti and M.-A. DePaoli, Synth. Met., 80(1996)263.

47. W. Li and M. Wan, Synth. Met., 92(1998)121.

48. W.A.Jr. Gazotti, R. Faez and M.-A. DePaoli, J. Ele.Chem., 415(1996)107.

49. Y. Cao and A.J. Heeger, Synth. Met., 52 (1992) 193.

50. N.S. Sariciftci, L. Smilowitz, Y. Cao and A.J. Heeger, Synth. Met., 55 (1993) 188.

51. J.E. Osterholm, Y. Cao, F. Klevetter and P. Smith, Synth. Met., 55 (1993) 1034.

52. C.Y. Yang, P. Smith, A.J. Heeger, Y. Cao and J.E. Osterholm, Polymer, 35 (1994) 1142.

53. T. Taka, J. Laakso and K. Levon, Sol. Stat. Comm., 92 (1994) 393.

54. J.M. Ko and I.J. Chang, Synth. Met., 68 (1995) 233.

55. T. Vikki and O.T. Ikkala, Synth. Met., 69 (1995) 235.

56. S.J. Davies, T.G. Ryan, C.J. Wilde and G. Bayer, Synth. Met., 69(1995)209.

57. B.Y. Choi, I.J. Chung, J.H. Chung and J.M. Ko, Synth. Met., 99(1999)253.

58. P. S. Rao, D. N. Sathyanarayana and T. Jeevananda, In Adv. Fun. Mol. & Poly., H. S. Nalwa (ed.), Gordon and Breach, Tokyo, 3 (2001)79.

59. G. E. Wnek, Synth. Met., 15, (1986) 213.

60 . B. Lundberg, W. R. Salaneck and I. Lundström, Synth. Met., 21(1987)143.

61. W. R. Salaneck, I. Lundström, T. Hjertberg, C. B. Duke, A. Paton, E. M. Conwell, W. S.Huang, N. L. D. Somasri, A. F. Richter and A. G. Mac Diarmid, Synth. Met., 18(1987) 311.

62. J. Margolis, Conductive Polymers and Plastics, Chapman and Hall, (1989)121.

63. Carter, C. Barry; Norton, M. Grant "Ceramic materials: science and engineering" Springer, (2007).

64. Shriver, D. F.; Atkins, P. W.; Overton, T. L.; Rourke, J. P.; Weller, M. T.; Armstrong, F. A. "Inorganic Chemistry" W. H. Freeman, New York, 2006.

65. R Skomski, J. Phys.: Condens. Matter 15, R1-R56, (2003).

66. T. Feried, G. Shemer, G. Markovich, Adv. Mater., 13 (2001) 1158.

67. M. Greenblatt, Chem, Rev. 88 (1988) 31.

68. J. D. Lee, Concise Inorganic Chemistry (IVEd.), ELBS, London (1997).

69. T. Siyama, A. Kato, K Fujiishi, M. Nagatani, A new detector for gaseous components using semi-conducting thin films, Anal. Chem. 34 (1962) 1502.

70. J. N. Taguichi, Jpn. Patent 45-38200 (1962).

71. J. N. Taguichi, UK Patent 1280809 (1970).

72. Mitrovics, J., Ulmer, H., Weimar, U., Göpel, W., Acc. Chem. Res. 31 (1998) 307.

73. Robert W, Cattrall, Chemical Sensors Oxford University Press (1997), New York

74. N. Golego, S.A. Studenikin and M. Cocivera, J. Electrochem. Soc.,147 (2000)1592.

75. Y. Lin, Z. Zhang, Z. Tang, F. Yuan and J. Li, Adv. Mater. Opt. Electron, 9, (1999)206.

76. X. Wang, J. Song, J. Liu and Z.L. Wang, Science, 316, (2007) 102.

77. K. Keren, R.S. Berman, E. Buchstab, U. Sivan, E. Braun, Science, 302, (2003)1380.

78. K. Nomura, H. Ohta, K. Ueda, T. Kamiya, M. Hirano, H. Hosono, Science, 300,(2003) 1269

79. T. Nakada, Y. Hirabayashi, T. Tokado, D. Ohmori, T. Mise, Sol. Energy, 77, (2004)739

80. S. Y. Lee, E. S. Shim, H. S. Kang, S. S. Pang, J. S. Kang, Thin Solid Films, 437, (2005) 31

81. R. Könenkamp, R. C. Word, C. Schlegel, Appl. Phys. Lett., 85, (2004) 6004

82. S. T. Mckinstry, P. Muralt, J. Electroceram., 12, (2004) 7

83. Z. L. Wang, X. Y. Kong, Y. Ding, P. Gao, W. L. Hughes, R. Yang,Y. Zhang, Adv. Funct. Mater., 14, (2004) 943

84. M. S. Wagh, L. A. Patil, T. Seth, D. P. Amalnerkar, Mater. Chem. Phys., 84, (2004) 228

85. Y. Ushio, M. Miyayama, and H. Yanagida, Sensor Actuat., B 17, (1994) 221

86. H. Harima, J. Phys. Condens. Matter., 16, (2004) 5653

87. S. J. Pearton, W. H. Heo, M. Ivill, D. P. Norton, and T. Steiner, Semicond. Sci. Technol., 19, (2004) 59

88. M.A. Garcia, J.M. Merino, E.F. Pinel, A.J. Quesada, D. Venta, M.L.R. Gonza, G.R. Castro, P. Crespo, J. Llopis, J.M.G. Calbet, A.Hernando, Nano letter., 7 (2007) 1489.

89. P. Crespo, J. Llopis, J.M.G. Calbet, A.Hernando, Nano letter., 7 , (2007) 1489

90. Q.P. Zhong , E. Matijevic, J.Mater. Chem., 3, (1996) 443

91. W. Lingna , M. Mamoun, J.Mater. Chem., 9 (1999) 2871

92. D.W. Bhnemann, C. Kormann, M .R. Hoffmann, J.Phys. Chem., 91(1987)3789

93. Z. Hui, Y.Deren, M. Xiangyang, J.Yujie, X. Jin, Nanotechnology., 15 , (2004) 622

94. J.Zhang, L. D. Sun, J.L. Yin, Chem.Mater., 14 , (2002) 4172.

95. W.J. Li, E.W. Shi, Z.W. Yin, J.Mater.Sci.Lett., 20, (2001) 1381

96. Forsyth J.B., Hull S., J. Phys. Condens. Matter, 3 (1991) 5257.

97. Jing Wang, Feng-Qing Wu, Kai-He Shi, Xiao-Hua Wang, Peng-Po Sun Sensors and Actuators B 99 (2004) 586.

98. Scott T. McGovern, Geoffrey M. Spinks, Gordon G. Wallance, Sensors and Actuators B, 10(2005)657.

99.Narsimha Parvatikar, Syed Khasim, M. Revansiddappa, Shilpa Jain, S V Bhoraskar, and M V N Ambika Prasad, Sensors and Actuators B, 114 (2006) 599.

100.Narsimha Parvatikar, Shilpa Jain, C M Kanamadi, B K Chougule, S V Bhoraskar, and M V N Ambika Prasad, Journal of app. Pol. Sci., 103(2007)653.

101.Labidi.A,Lambert-Mauriat.C, Jacolin.C, Bendahan.M, Maaref Aguir.K, Sensors and Actuators B 119 (2006) 374.

102. B.Y. Wei, M.C. Hsu, P.G. Su, H.M. Lin, R.J. Wu, H.J. Lai, Sensors and Actuators, B,101 (2004) 81.

103.Kida.Tetsuya, Kawasaki.Kohsuke, Lemura.Kaoru, Teshima Kazusha, Nagano Masamitsu, Sensors and Actuators B 119 (2006) 562.

104.Barazzouk. S, Tandon.R.P, Hotchandani.S ., Sensors and Actuators B 119 (2006) 691.

105. Abhilash Srivatsav, S.T.Lakshmikumar, A.K.Srivatsav,Rashmi and Kiran Jain, Sensors and Actuator B 126(2) (2007) 583.

106. Q.Y. Soundararajaha, B.S.B. Karunaratnea, R.M.G. Rajapakseb, Materials Chemistry and Physics 113 (2009) 850.

107. Peng Liu et.al.,Current Opinion in Solid State and Materials Science 12 (2009) 9.

108. A. L. Kukla, A. S. Pavluchenko, Yu. M. Shirshov, N.V. Konoshchuk and O. Yu. Posudievsky., Sensors and Actuators B 135(2)(2009)541.

CHAPTER – II

2.0. Introduction

Polyaniline (PANI) exists in a variety of forms that differ in chemical and physical properties. The most common green protonated emeraldine has conductivity on a semiconductor level of the order of 100 S cm^{-1}, many orders of magnitude higher than that of common polymers (<10^{-9} S cm^{-1}) but lower than that of typical metals (>10^4 S cm^{-1}). Protonated PANI, (e.g., PANI hydrochloride) converts to a non-conducting blue emeraldine base when treated with ammoniumhydroxide [1–4] (Fig2.1).

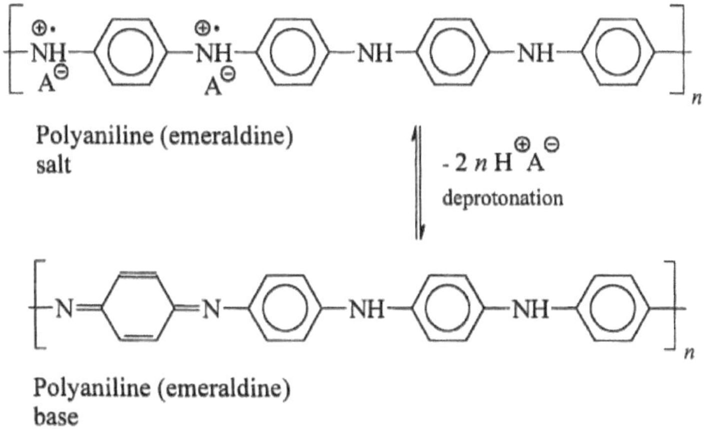

Figure 2.1. Polyaniline (emeraldine) salt is deprotonated in the alkaline medium to polyaniline (emeraldine) base. A– is an arbitrary anion, e.g., chloride.

The changes in physicochemical properties of PANI occurring in the response to various external stimuli are used in various applications [5–6], e.g., in organic electrodes, sensors, and actuators [7–9]. Other uses are based on the combination of electrical properties typical of semiconductors with materials parameters characteristic of polymers, like the development of "plastic" microelectronics [10], electrochromic devices [11], tailor-made composite systems [12,13], and "smart" fabrics [14]. The establishment of the physical properties of PANI reflecting the conditions of preparation is thus of fundamental importance.

The efficient polymerization of aniline is achieved only in an acidic medium, where aniline exists as an anilinium cation. A variety of inorganic and organic acids of different concentration have been used in the syntheses of PANI; the resulting PANI, protonated with various acids, differs in solubility, conductivity and stability. For the present study, we have selected hydrochloric acid in equimolar proportion to aniline, i.e., aniline hydrochloride was used as a monomer. The handling of solid aniline salt is preferred to liquid aniline from the point of view of toxic hazards. Peroxydisulfate is the most commonly used oxidant and its ammonium salt was preferred to the potassium counterpart because of its better solubility in water. The concentration of aniline hydrochloride was set to 0.2 M. various oxidant/ monomer ratios have been used in the literature. To minimize the presence of residual aniline and to obtain the best yield of PANI, the stoichiometric peroxydisulfate/aniline ratio 1.25 is recommended [15] (Fig. 2). The polymerization is completed within 10-15 min at room temperature and within 1 h at 0–2 °C [16]. The oxidation of aniline is exothermic so the temperature of the reaction mixture can be used to monitor the progress of reaction [17–18]. Temperature profile is well reproducible.

$$4\,n \!\langle\bigcirc\rangle\!-\!NH_2.HCl + 5\,n\,(NH_4)_2S_2O_8 \longrightarrow$$

$$\left[-NH\!-\!\langle\bigcirc\rangle\!-\!\overset{\oplus\cdot}{\underset{Cl^{\ominus}}{NH}}\!-\!\langle\bigcirc\rangle\!-\!NH\!-\!\langle\bigcirc\rangle\!-\!\overset{\oplus\cdot}{\underset{Cl^{\ominus}}{NH}}\!-\!\langle\bigcirc\rangle\!-\right]_n$$

$$+\,2\,n\,HCl +\,5\,n\,H_2SO_4 +\,5\,n\,(NH_4)_2SO_4$$

Figure 2.2. Oxidation of aniline hydrochloride with ammonium peroxydisulfate yields polyaniline (emeraldine) hydrochloride.

2.1. Chemical Synthesis of Emeraldine Salt

The synthesis was based on mixing aqueous solutions of aniline hydrochloride and ammonium peroxydisulfate (Fig. 2.2) at room temperature, followed by the separation of PANI hydrochloride precipitate by filtration and drying. Aniline hydrochloride (equi molar volumes of aniline and hydrochloric acid) was dissolved in distilled water in a volumetric flask to 100 ml of solution. Ammonium peroxydisulfate (0.25M) was dissolved in water also to 100 ml of solution. Both solutions were kept for 1 hour at room temperature (25°C), then mixed in a beaker, stirred with a mechanical stirrer, and left at rest to polymerize. Next day, the PANI precipitate was collected on a filter, washed with 300-mL portions of 0.2 M HCl, and similarly with acetone. Polyaniline (emeraldine) hydrochloride powder was dried in air and then in vacuum at 60°C. Polyanilines prepared under these reaction and processing conditions are further referred to as "standard" samples. Additional polymerizations were carried out in an ice bath at 0–2°C. The acidity of the reaction mixture was increased by replacing 10 ml of water with 10 ml of 2 M HCl. The flow chart showing the various steps involved in chemical synthesis of polyaniline is as shown in figure 2.3.

2.2. Synthesis of Polyaniline and its Composites

All Chemicals used were analytical reagent (AR) grade. The monomer aniline was doubly distilled prior to use. Ammonium persulphate $((NH_4)_2S_2O_8)$, Hydrochloric acid (HCl), Zinc ferrite $(ZnFe_2O_4)$, Nickel zinc oxide $(NiZnO_3)$ and Copper oxide (CuO) were procured from sigma and were used as received.

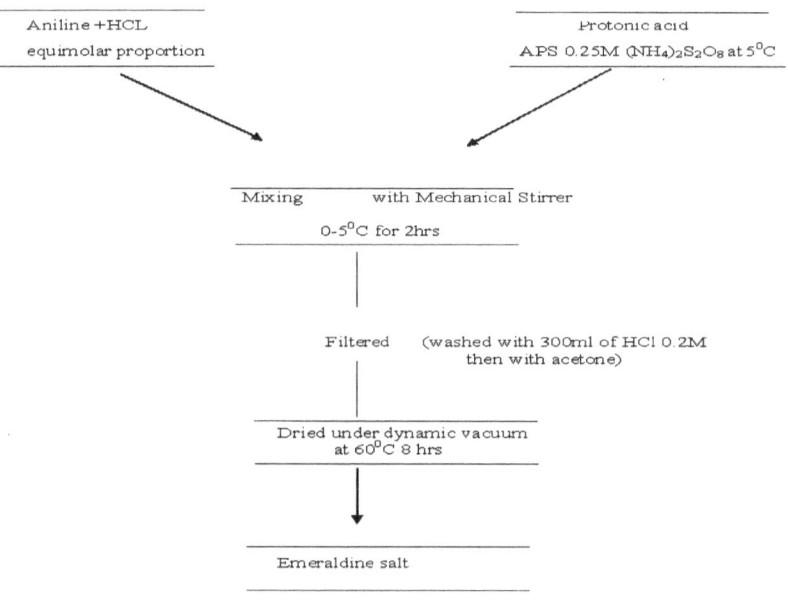

Figure 2.3 shows the flow chart showing the various steps involved in chemical synthesis of polyaniline.

2.2.1. Synthesis of Polyaniline / $ZnFe_2O_4$ Composites

0.1 mole aniline monomer is dissolved in 1 mole hydrochloric acid to form polyaniline hydrochloride. Fine graded pre-sintered $ZnFe_2O_4$ (AR grade, SD-Fine Chem.) powder in the weight percentages (wt %) of 10, 20, 30, 40 and 50 is added to the polymerization mixture with vigorous stirring in order to keep the $ZnFe_2O_4$ powder suspended in the solution. To this reaction mixture, APS as an oxidant is added slowly with continuous stirring for the period of 4 hrs at temperature 5^0C. Polymerization of aniline takes place over fine grade zinc ferrite particles. The resulting precipitate is filtered and washed with distilled water until the filtrate becomes colorless. Acetone is used to dissolve any unreacted aniline. After washing, the precipitate is dried under dynamic vacuum at 60^0C for 24 hrs to get resulting composites [19]. In this way five different polyaniline/ $ZnFe_2O_4$ composites with different weight percentage of zinc ferrite (10, 20, 30, 40 and 50) in polyaniline have been synthesized. All the composites are crushed into fine powder in an agate mortar in the presence of acetone medium.

2.2.2. Synthesis of Polyaniline/Ni ZnO_3 Composites

Polyaniline and Polyaniline/NiZnO$_3$ composites are prepared by oxidative polymerization method. The equimolar volume of aniline and hydrochloric acid is mixed and stirred for 20min to form aniline hydrochloride. Fine graded pre-sintered NiZnO$_3$(AR grade, Fine Chem.) powder in the weight percentages (wt %) of 10, 20, 30, 40 and 50) is added to the polymerization mixture with vigorous stirring in order to keep the NiZnO$_3$ powder suspended in the solution. To this reaction mixture, $[(NH_4)_2S_2O_8)]$ used as an oxidant, is added slowly drop-wise with continuous stirring for the period of 4 hrs at temperature 5^0C. Polymerization of aniline takes place over fine grade Nickel zinc oxide particles. The resulting precipitate is filtered under suction and washed with distilled water until the filtrate becomes colorless. Acetone is used to dissolve any unreacted aniline. After washing, the precipitate is dried under dynamic vacuum at 60^0C for 24 hrs to get resulting composites [20]. In these way five different Polyaniline/Ni ZnO_3 composites with different weight percentage of Nickel zinc oxides (10, 20, 30, 40 and 50 wt. %) in Polyaniline has been synthesized. All the composites are crushed into fine powder in an agate mortar in the presence of acetone medium.

2.2.3. Synthesis of Polyaniline – CuO Composites

0.1 mol of aniline was dissolved in 1 M HCl to form aniline hydrochloride. Copper oxide is added in the weight percent of 10, 20, 30, 40 and 50 to aniline hydrochloride solution with vigorous stirring in order to keep the tantalum pentaoxide suspended in the solution. To this reaction mixture, 0.1 M of ammonium persulphate $[(NH_4)_2S_2O_8]$ which acts as the oxidant was added slowly with continuous stirring for 4 – 6 hours at $0 – 5 ^0$ C. The precipitated powder recover ware vacuum filtered and washed with water and acetone. Finally the resultant precipitate was dried in an oven for 24 hours to achieve a constant weight. In this way 5 different Polyaniline / CuO composites with different wt % of CuO (10, 20, 30, 40 and 50) in polyaniline have been synthesized [21]. All the composites so obtained above are crushed into fine powder in an agate mortar in the presence of acetone medium.

2.3. Preparation of Pellets

The powders of polyaniline, polyaniline – $ZnFe_2O_4$, polyaniline – Ni ZnO_3 and polyaniline – CuO composites so obtained from synthesis techniques discussed in the early sections are crushed and ground finely in the presence of acetone medium in agate mortar. This powder is pressed to form pellets of 10 mm diameter and thickness which varies from 1 to 2 mm by applying pressure of 90 MPa in a hydraulic press.The pellets of polyaniline and its composites so obtained from above mentioned techniques are coated with silver paste on either side of the surfaces. The copper electrodes are placed on each of the surface to obtain better contacts.

References

1. A. G. MacDiarmid and A. J. Epstein. Faraday Discuss. Chem. Soc.88 (1989)317.

2. J. Stejskal, P. Kratochvil, A. D. Jenkins. Polymer 37 (1996) 367.

3. D. C. Trivedi. Handbook of Organic Conductive Molecules and Polymers, H. S. Nalwa (Ed.) Vol. 2 (1997) 505, Wiley, Chichester.

4. Than-Dinh Nguyen, Cao-Thang-Dinh and Trong-On Do, Nanoscale, 3 (2011) 1861

5. B. G. Levi. Phys. Today 53 (12) (2000) 19.

6. A. G. MacDiarmid. Angew. Chem., Int. Ed. 40, (2001) 2581.

7. Z. Jin, Y. Su, Y. Duan. Sensor & Actuators B 72 (2001) 75.

8. P. T. Sotomayor, I. M. Raimundo, Jr., A. J. G. Zarbin, J. J. R. Rohwedder, G. O. Netto, O. L. Alves. Sensor & Actuators B 74 (2001) 157.

9. L. A. P. Kane-Maguire and G. G. Wallace. Synth. Met., 39 (2001) 119.

10. R. J. Hamers. Nature 412 (2001) 489.

11. U.Jeong, X.Teng, Y.Wang.H.Yang, Y.Xia, Adv. Mater;19(2007)33.

12. A.H.Lu, E.L.Salabas, F.Schueth, Angew. Chem., Int.Ed, 46(2007)1222.

13.G. K. Elyashevich, L. Terlemezyan, I. S. Kuryndin, V. K. Lavrentyev, P. Mokreva, E.Yu. Rosova, Yu. N. Sazanov. Thermochim. Acta, 23 (2001)374.

14. Aashis S. Roy, Koppalkar R.Anilkumar, M.V.N.Ambika Prasad; J.App. Poly.Sci.;121(2011)675.

15. S. P. Armes and J. F. Miller. Synth. Met. 22 (1988) 385.

16. T. Sulimenko, J. Stejskal, I. Krivka, J. Prokes. Eur. Polym. J. 37 (2001)219.

17. Y. Fu and R. L. Elsenbaumer. Chem. Mater. 6 (1994) 671.

18. Igwe.H.U., Ugwu.E.I., Adv.in Appl Sci.Res,1(2010)240.

19. Ramesh Patil, Aashis S. Roy, Koppalkar R. Anilkumar, Srikant Ekhelikar; J. App. Poly. Sci, vol. xxx (2011) 262

20. Aashis S. Roy, Koppalkar R. Anilkumar, M. V. N. Ambika Prasad, J. App. Poly. Sci, Vol.xxx, (2011); DOI 10.1002/app.34696

21. S.Radhakrishnan, R.Muthukannan, U.Kamatchi, Chepuri.R.K.Rao and M.Vijayan, Indian Journal of Chemistry; 50A, (2011) 970.

Chapter III

3.0. Introduction

X-Ray diffraction (XRD) Studies

X – ray spectroscopy is one of the important and prominent tools for the study of structure of molecules and crystals. There are certain advantages in the use of X – ray methods as against the optical ones. X – rays have intrinsic characteristics possessing more energy than that of light as such these can easily penetrate through any opaque media. This very property will give us an interference pattern of coherent beams being diffracted from the true crystal lattice of the matter under investigation. There are many X–ray methods for the study of the crystals. But amongst them, the most useful and powerful techniques are interferometry and diffractometry. With the application of these methods, one can obtain information such as the deformation of the crystals, electronic energy density in terms of a function of width [1-4].

3.1. X-Ray Diffraction

Principal Methods of Structure Elucidation:

The X-ray structure analysis is based on the phenomenon of diffraction of X-rays by a substance. X-rays are short electromagnetic waves. In the X-ray diffraction analysis, wavelengths of 0.5-2.5A° are used. The X-ray study give information about the nature and structure of the materials and also helps to know the phase present in the end product of preparation. Analysis of the X-ray data gives the values of the unit cell parameters of the system [5-7]. The slow disappearance of certain peaks and emergence of new peaks in the pattern reveals the associated structural phase changes occurring due to the variation in temperature, dopant concentration etc.

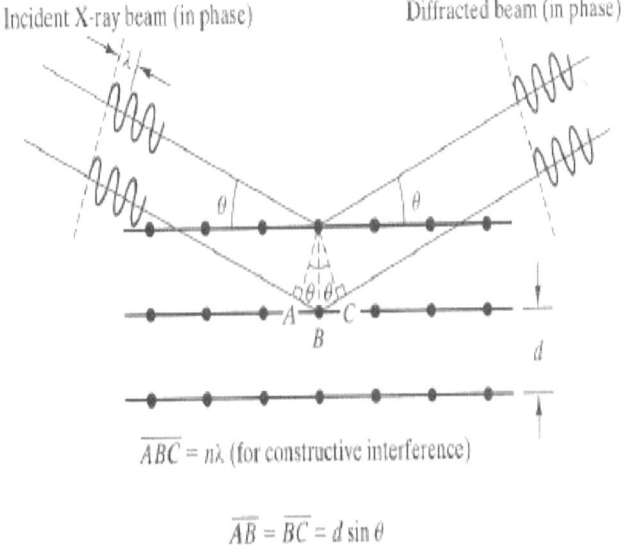

$$\overline{ABC} = n\lambda \text{ (for constructive interference)}$$

$$\overline{AB} = \overline{BC} = d \sin \theta$$

Figure-3.1. the reflection of X-rays by a family of atomic planes

When a monochromatic X-ray falls on a crystal, whose dimensions are of the same order of magnitude as the wavelength as shown in figure 3.1 then diffraction occurs according to Braggs law given by,

$$2d \sin \theta = n\lambda$$

where d is interplanar spacing, θ is glancing angle, λ is wave length of X-ray and n is the order. The 2θ values obtained from the position of the Bragg peaks from the XRD patterns are used to obtain the interplanar spacing expressed in terms of Miller indices (hkl), are substituted into the above equation and a number of relations which are useful for indexing and calculating the lattice parameter are obtained. The distance between the planes (d) may be found by substituting the value of lattice parameter (a) and (hkl) in the following equations:

1. Cubic system:

$$d_{hkl} = \frac{a}{\sqrt{h^2 + k^2 + l^2}}$$

2. Orthorhombic system

$$d_{hkl} = \frac{1}{\sqrt{\frac{h^2}{a^2} + \frac{k^2}{b^2} + \frac{l^2}{c^2}}}$$

3. Tetragonal system:

$$d_{hkl} = \frac{a}{\sqrt{h^2 + k^2 + l^2 \left(\frac{a}{c}\right)^2}}$$

For the monoclinic and triclinic systems, which are most typical of a number of polymers, the corresponding formulas have a more complex form.

The intensity and direction of X-rays that have been diffracted by a crystal are recorded by a quantum detector (a Geiger counter or a scintillation counter) or by a photographic technique.

In order to elucidate the structure of crystals, various methods are used [8, 9]. These methods include Laue's method, the rotating crystal method and the Debye-Scherrer method (the powder technique).

3.2. Powder method

In this method a monochromatic beam of X-rays is made to strike a fine grained specimen in a thin walled capillary tube. The diffraction occurs simultaneously from individual crystallites that are oriented with planes having some incident glancing angle (θ) satisfying Bragg's law. In diffractometers the sample is placed at the center and diffracted X-rays are recorded by proportional counter. When the sample is rotated through an angle θ, the proportional counter is rotated through 2θ. The diffractometer gives the variation of intensity of diffraction lines with diffraction angle 2θ.

3.3. Experimental Technique

The X-ray diffraction patterns of the samples in this present study are obtained on Philips X-ray diffractometer shown in figure 3.1.1 using CuK_α radiation (λ = 1.5406 Å). The diffractograms were recorded in terms of 2θ in the range $20^0 - 120^0$ with a scanning rate of 2^0 per minute.

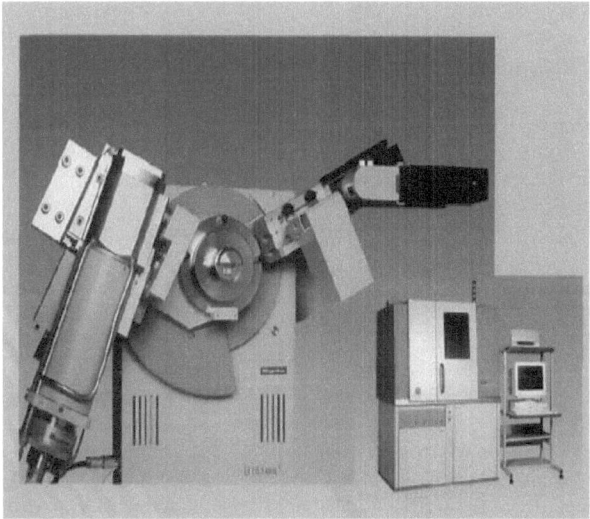

Figure 3.1.1 shows the image of X-ray diffractometer

3.4. Results and Discussion

3.4.1. Polyaniline

Figure 3.1.2 Shows X-ray diffraction pattern of Polyaniline. Careful analysis of X-ray diffraction of polyaniline suggests that it has amorphous nature with a broad peak centered on 2θ ≈ 26.40^0.

3.4.2. Polyaniline / ZnFe₂O₄ composites

Figure 3.1.3 shows the X-ray diffraction pattern of Polyaniline – $ZnFe_2O_4$ composite with 50 wt % of $ZnFe_2O_4$ in polyaniline. It is seen from figure that the peak of $ZnFe_2O_4$ indicates the crystalline nature of the composite. By comparing the XRD pattern of composite with that of PANI, the prominent peaks corresponds to 2 θ = 18.59⁰, 30.09⁰, 35.55⁰, 56.43⁰ and 62.32⁰ are due to (220), (311), (400) (422) and (333) planes of $ZnFe_2O_4$. By comparing the XRD patterns of the composite and PANI, it is confirmed that $ZnFe_2O_4$ (JCPDS 06-0696) has retained its structure even though it is dispersed in PANI during polymerization reaction [10].

3. 4.3. Polyaniline / NiZnO₃ composites

Figure 3.1.4 shows the X-ray diffraction pattern of Polyaniline – $NiZnO_3$ composite with 50 wt % of $NiZnO_3$ in polyaniline. It is seen from the figure that, the peaks of $NiZnO_3$ indicates the crystalline nature of the composite. By comparing the XRD pattern of composite with that of PANI, the prominent peaks corresponding to 2 θ = 18.02⁰, 24⁰, 26.5⁰, 30.7⁰ and 61.02⁰, are due to (100), (002), (101), (102) and (110) planes of $NiZnO_3$ (JCPDS card No. 04-0835 and No.05-0664). By comparing the XRD patterns of the composite, it is confirmed that $NiZnO_3$ has retained its structure even though it is   dispersed in PANI during the polymerization reaction.

3. 4.4. Polyaniline / CuO composites

The X-ray diffraction pattern of PANI/CuO composite is shown in figure(3.1.5). There are five prominent and two less prominent peaks representing Bragg's reflections of copper oxide. While the broad peaks representing the periodicity parallel to the polymer chains of PANI was not clearly observed at the 2θ value of 26.5 due to weak intensity in comparison to the copper oxide peak (JCPDS No.05-661). Results from both FTIR and X-ray diffraction measurements have provided additional evidence that the polymerization of PANI has been successfully obtained on the surface of the CuO composites.

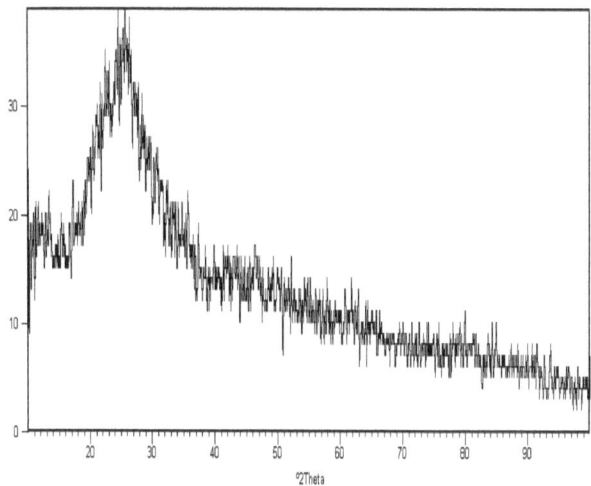

Figure 3.1.2. X-ray diffraction pattern of polyaniline

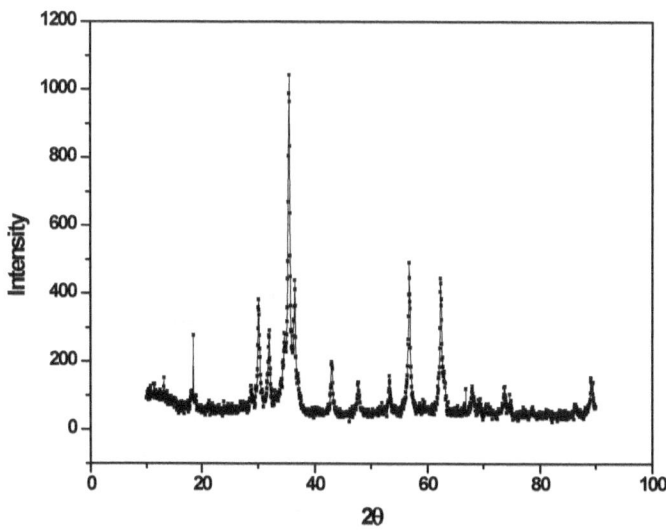

Figure 3.1.3. X-ray diffraction pattern of Polyaniline – $ZnFe_2O_4$ composites

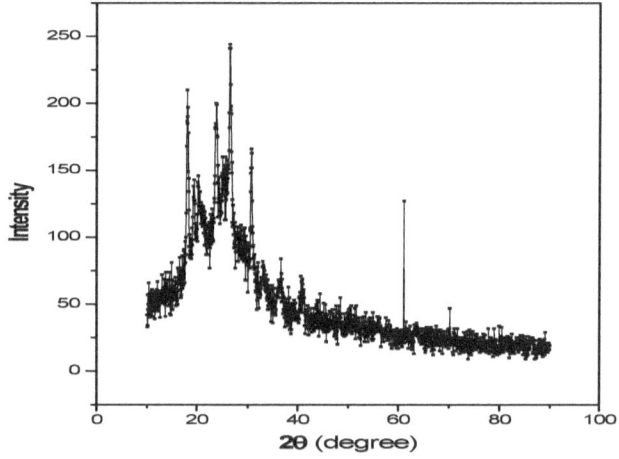

Figure 3.1.4.X-ray diffraction pattern of Polyaniline – $NiZnO_3$ composites

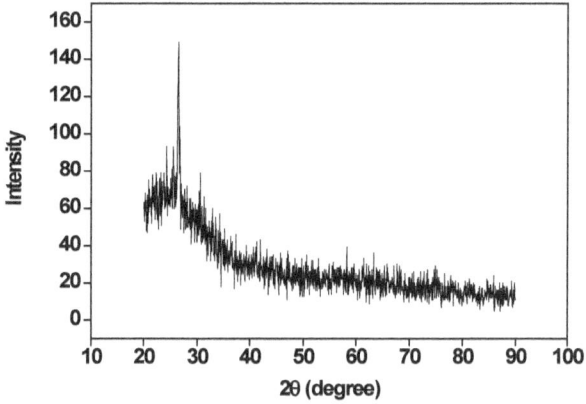

Figure 3.1.5. X-ray diffraction pattern of Polyaniline – CuO composite

Average inter chain separation can be estimated from these maxima using the relation

$$R = \frac{5}{8}\left[\frac{\lambda}{\sin\theta}\right]$$

[11-14] where λ is the x-ray wavelength of CuKα and θ is the diffraction angle at the maximum intensity.

Dopant (Primary /Secondary)	2θ	Inter Chain separation (A°) $R = \frac{5}{8}\left[\frac{\lambda}{\sin\theta}\right]$
HCl	25.40	2.24
$ZnFe_2O_4$	34.34	1.706
NiZnO	28.66	2.007
CuO	25.25	2.257

3.5. Introduction Infra Red Spectroscopy

The electrical and magnetic properties of materials depend upon their chemical composition, cation distribution and method of preparation. The vibrational, electronic and magnetic dipoles spectra can give information about the position and valance of the ions in the crystal lattice. The infrared spectrum is an important undestructive tool to describe the various ordering problems.

3.5.1. Experimental technique

The IR spectra of all the samples are recorded on Perkin Elmer (model 783) IR spectrometer in KBr medium at room temperature. For recording IR spectra, powders are mixed with KBr in the ratio 1:25 by weight to ensure uniform dispersion in KBr pellets. The mixed powders are pressed in a cylindrical die to obtain clean discs of approximately 1 mm thickness.

The characterization of polyaniline and its composites by spectroscopic methods is important, as it gives information not only about various molecular – levels interactions but also on the type of charge carriers.

Infrared spectroscopy is a powerful tool to determine the structural changes that occurs during doping and dedoping process. Various groups have reported IR results of polyaniline, but the IR results of composites are scarce.

3.5.2. Results and Discussions

The occurrence of various bands in IR spectra may be attributed due to the following factors.

Frequency (cm⁻¹)	Observed band due to
1610, 1593 and 1503	C = C stretching
1550 and 1500	C = N stretching
1040, 960 and 790	C – N stretching
1050	N – H bending deformation
726 and 1326	C – H bending deformation

3.5.3. Polyaniline

Figure 3.2.0 shows the FTIR spectra for pure Polyaniline. The characteristic absorption peaks are found to be at 2922 cm⁻¹ is due to the C-H stretching, 1566 cm⁻¹ corresponds to C = C stretching vibration of quinoid ring, 1548 cm⁻¹ for C=N bond stretching, 1494 cm⁻¹ corresponds to stretching vibration of benzenoid ring, 1406 cm⁻¹ is the characteristic vibration mode of C - H bonding of aromatic nuclei, 1302 cm⁻¹

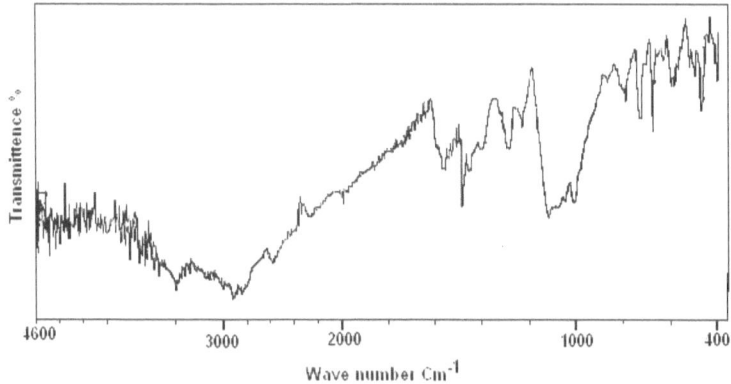

Figure 3.2.0 shows the FTIR spectra for pure Polyaniline.

assigned to the stretching of C-N bonds of aromatic amines, 1140 cm^{-1} an strong band which considered to be the measure of degree of electron delocalization 796cm^{-1} corresponding to the N-H out of plane bending in rocking mode.734 cm^{-1} and 684 cm^{-1} are due to the out of plan blending of C-H bond in aromatic ring, respectively.

3.5.4. Polyaniline / ZnFe$_2$O$_4$ composites

Figure 3.2.1(a-f) shows the FTIR spectra of polyaniline/ZnFe$_2$O$_4$ composites at different percentage (10, 20, 30, 40 and 50 wt. %). The absorption peaks are found to be at 3441cm^{-1} correspond for N-H stretching vibration, 1581 cm^{-1} corresponds to C = C stretching vibration of quinoid ring, 1481 cm^{-1} corresponds to stretching vibration of benzenoid ring, 1300 cm^{-1} is for the stretching of C-N bonds of aromatic amines, 1240 cm^{-1} for the C-N stretching of benzenoid ring, 1140 to 1145 cm^{-1}corresponds to C-H in plane of aromatic rings found to be an strong band which considered to be the measure of degree of electron delocalization, and other bands like 802 cm^{-1},738 cm^{-1}, and 686 cm^{-1} are due to the out of plan blending of C-H bond in aromatic ring, respectively. The samples show characteristic absorptions bands of Zinc ferrite, the absorptions bands at 507 cm^{-1} is due to the intrinsic vibration of Zn^{+2} which is present in tetrahedral positions and around 415 cm^{-1} which corresponds to the vibration of octahedral group of $Fe^{+3}O^{-2}$, which confirm the formation of the polyaniline / Zinc ferrite composites [15].

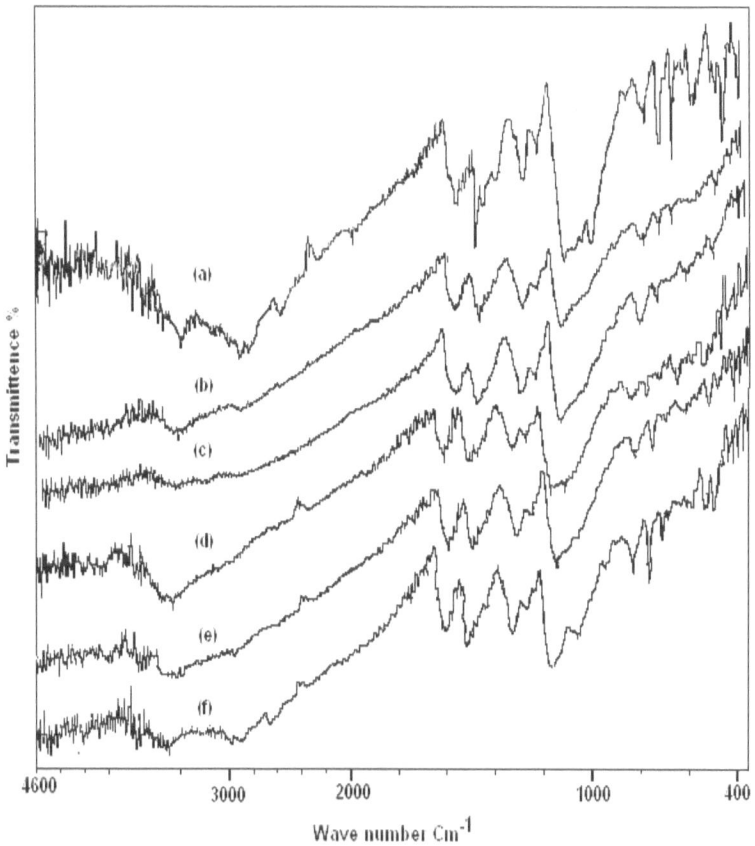

Figure 3.2.1(a-f) shows the FTIR spectra of PANI/ZnFe$_2$O$_4$ composites of different weight percentage.

3.5.5.. Polyaniline / NiZnO$_3$ composites

Figure 3.2.2 (a-f) shows that the FTIR spectra of PANI/NiZnO$_3$ composites of 10 wt %. The characteristic absorption peaks is found to be at 3433 cm^{-1}, 1556 cm^{-1}, 1487 cm^{-1}, 1302 cm^{-1}, 1236 cm^{-1}, 1140 cm^{-1}, 800 cm^{-1}, 738 cm^{-1}, 584 cm^{-1}, 509 cm^{-1} & 422cm^{-1} respectively. The peaks at 3433cm^{-1} is corresponds to N-H stretching vibration 1556 cm^{-1} is for stretching vibration of C=C bonds of quiniod rings,

59

1487cm^{-1} is corresponds to C=C stretching vibration of benzoniod ring, 1302cm^{-1} is due to the stretching of C-N bonds of aromatic amines 1236cm^{-1} corresponds to C-N stretching of benzoniod ring 1140cm^{-1} corresponds to C-H in plane of aromatic rings is found to be a strong band which is considered to be the measure of degree of electron delocalization $800 \text{cm}^{-1}, 738 \text{cm}^{-1}$ & 584cm^{-1} are due to the out of plane bending of C-H bond in aromatic ring. The composites also show the absorption bands at 509cm^{-1} is due to the intrinsic vibration of Ni^{+2} which is present in tetrahedral position and around 422cm^{-1} is vibration of octahedral group of Zinc which conform the formation of $PANI/NiZnO_3$ composites.

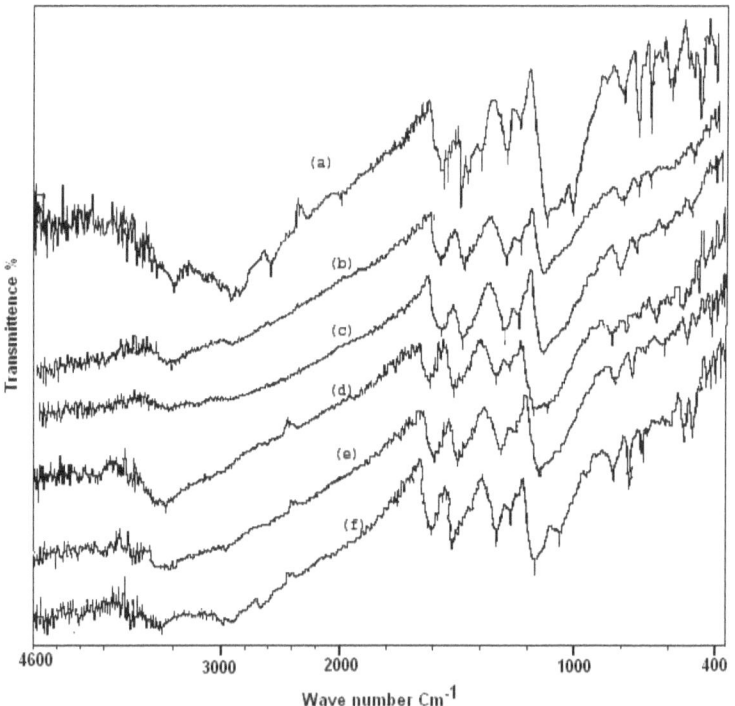

Figure 3.2.2 (a-f) shows the FTIR spectra of $PANI/NiZnO_3$ composites of different weight percentage.

3.5.6. Polyaniline / CuO composites

Figure 3.2.3(a-f) shows the FTIR spectra of PANI and PANI/CuO composites of different weight percentage (10, 20, 30, 40 and 50 wt%). The FTIR spectra (figure-a) show the pure PANI. The characteristic important peaks are found at 1566 cm^{-1} corresponds to C = C stretching vibration of quinoid ring, 1548 cm^{-1} for C=N bond stretching, 1494 cm^{-1} corresponds to stretching vibration of benzenoid ring, 1406 cm^{-1} is the characteristic vibration mode of C - H bonding of aromatic nuclei, 1302 cm^{-1} assigned to the stretching of C-N bonds of aromatic amines, 1140 cm^{-1} an strong band which considered to be the measure of degree of electron delocalization, 796 cm^{-1} corresponding to the N-H out of plane bending in rocking mode respectively.

The important peaks of PANI/CuO composites (figure- b) of different weight percentage (10, 20, 30, 40 & 50 wt%) are found to be in between 1572-1552 cm^{-1} is due to the C = C stretching vibration of quinoid ring , 1460 -1481 cm^{-1} corresponds to stretching vibration of benzenoid ring, 1246-1304 cm^{-1} is assigned for stretching of C-N bonds of aromatic amines, 1118-1143 cm^{-1} is due to a strong band which considered to be the measure of degree of electron delocalization and 501-599 cm^{-1} is due to the stretching vibration of M-O in composites.

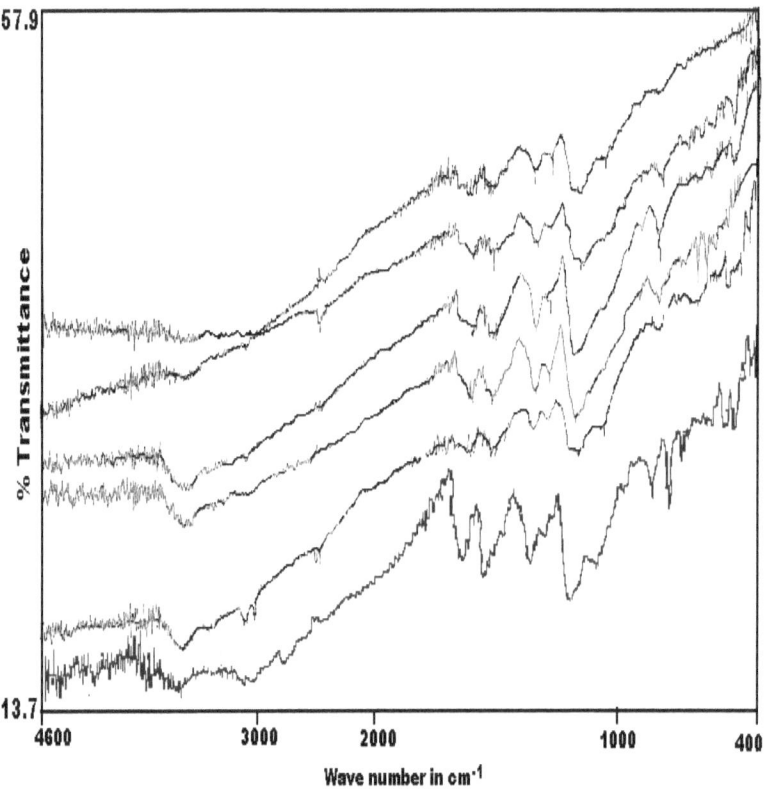

Figure 3.2.3 (a-f) shows the FTIR spectra of PANI/CuO composites of different weight percentage.

3.6. Introduction to Scanning Electron Microscopy (SEM)

The Micro structural studies on material provide information about:

1. Grain size of specimen

2. Amount of interfacial area per unit volume

3. Dimensions of constituent phases

4. Amount of Distribution of phase

5. Dislocation density

6. Volume fraction of precipitate and

7. Surface to volume ratio etc.

Some of the techniques employed for microstructure analysis are scanning electron microscopy, optical microscopy, field ion microscopy, field emission microscopy, X-ray microscopy and electron microprobe analysis. Of these, SEM is an extremely versatile technique for providing structural information over a wide range of magnifications, with an added advantage of depth of focus ($\sim$100 – 200 Å). The SEM has provided its potentiality to reveal the aspects of grain size, shape and orientation of pores, inclusions etc.

3.6.1. Experimental Technique

The powder morphology of polyaniline and its composites sintered in the form of pellets (to measure grain size) are investigated using Phillips XL30 ESEM scanning electronic microscope (SEM). The samples in the form of pellets are mounted on an aluminum platform, where conducting gold is sputtered on the sample to avoid charging at the sample surfaces. They were examined under SEM and selected areas were photographed.

3.6.2. Results and Discussions

SEM technique is applied primarily for the visualization of the sample surfaces, especially for the study of surface morphology, domains, pin hole defects and patterns. The images are formed by the interaction of electrons with samples in vacuum. The scanning electron micrographs of the samples in a system of polyaniline and its composites are presented in figures 3.3.0 to 3.3.3.

3.6.3. Polyaniline

SEM micrograph of conducting polyaniline synthesized by chemical oxidative method is shown in figure 3.3.0. It can be clearly seen that the micrograph of polyaniline is branched and homogeneous. Since Hydrochloric acid is used as protonic acid in the preparation of polyaniline, the presence of microcrystalline structure can be seen. The presence of microcrystalline structures in polyaniline in these particular samples can be confirmed from XRD studies. Since conducting polymers are very sensitive to the temperature, due to the interaction between electron and the sample, considerable amount of heat is generated which causes the development of mall crackening in the sample during SEM recording. A granular morphology of the microcrystalline structures is measured and is found to be about 312 nm in diameter for polyaniline which is consistent with other reports [16]. The contrast in the image is a result of differences in scattering from different areas of the surface as a result of geometrical differences.

Figure 3.3.0showes SEM image of pure PANI

3.6.4. Polyaniline / ZnFe₂O₄ composites

Figure3.3.1.(b) shows the Scanning Electron Micrograph of 10 wt. % of PANI/$ZnFe_2O_4$ composite where highly agglomerated cube like structure are seen. The crystallinity of the $ZnFe_2O_4$ decreases with the addition of PANI in it. It is found that zinc ferrite particles were homogeneously distributed throughout polyaniline matrix. The average grain size is found to be 600 nm.

PANI/ZnFe$_2$O$_4$ composite of 20 wt. % is shown in figure3.3.1.(c). The composite is highly clustered, spherical in shape and have interlinked to each other. The decrease in the inter-granular distance between the grains helps in charge transfer mechanism. The average grain size is found to be 1.5μm.

Figure3.3.1.(d) show the 30 wt. % of PANI/ZnFe$_2$O$_4$ composite. The images shows a highly crystalline granular flake like networking structure arranged in soccer shape and is well interlinked between each other. The average grain size is found to be 230 nm to 340 nm.

Figure3.3.1.(e) show that the 40 wt. % of PANI/ZnFe$_2$O$_4$ composite which is highly agglomerated and spherical in shape of about 0.5μm in granular size.

PANI/ZnFe$_2$O$_4$ composite of 50 wt. % is shown in figure 3.3.1..(f). It is clearly seen that the ferrites particles are not well bonded with the polyaniline due to increasing in the percolation limit to the ratio of filler concentration of the matrix.

From the figure3.3.1(a to f), it is found that, there is lots of change in the morphology of various wt% of ZnFe$_2$O$_4$ in PANI matrix's. The changes in the morphology were favorable for the transport mechanism in PANI / ZnFe$_2$O$_4$composites [17].

Figure3.3.1(a-f) shows the SEM image of PANI and PANI/ZnFe₂O₄ composites of different weight percentage (10, 20, 30, 40 and 50 wt%)

3.6.5. Polyaniline / NiZnO₃ composites

Figure 3.3.2.(b) shows that scanning electron micrograph of 10wt % of PANI/NiZnO₃ composites were highly agglomerated, irregular in shape and non porous in nature. The crystalline of PANI increase with addition of NiZnO₃. It is clearly observed in the micrograph cavity is formed; this may be due to the Ni^{+2} which gain the electron from polymer chain. The average grain size is found to be 2 μm.

Figure3.3.2(c) shows that the SEM image of 20wt % of $PANI/NiZnO_3$ composites. The grains are highly clustered, porous and spherical in shape. The average grain size is found to be 1.3μm.

Figure3.3.2.(d) Shows that the SEM image of 30 wt % of $PANI/NiZnO_3$ composites. The composites are formed a flake like structure with an average grain size of about 700 to 830nm.

Figure3.3.2.(e) shows that the SEM image of 40wt % of $PANI/NiZnO_3$ composites. The grins are clustered, spherical in shape. The composites matrix was found to be porous and average grain size is found to be about 1.7μm.

Figure3.3.2. (f) shows that the SEM image of 50wt % of $PANI/NiZnO_3$ composites. The particles are arranged in flake like, as well as some are spherical in shape. It is observed that the two flakes are connected in between by granular shaped particles. The average size is found to be 1.3μm.

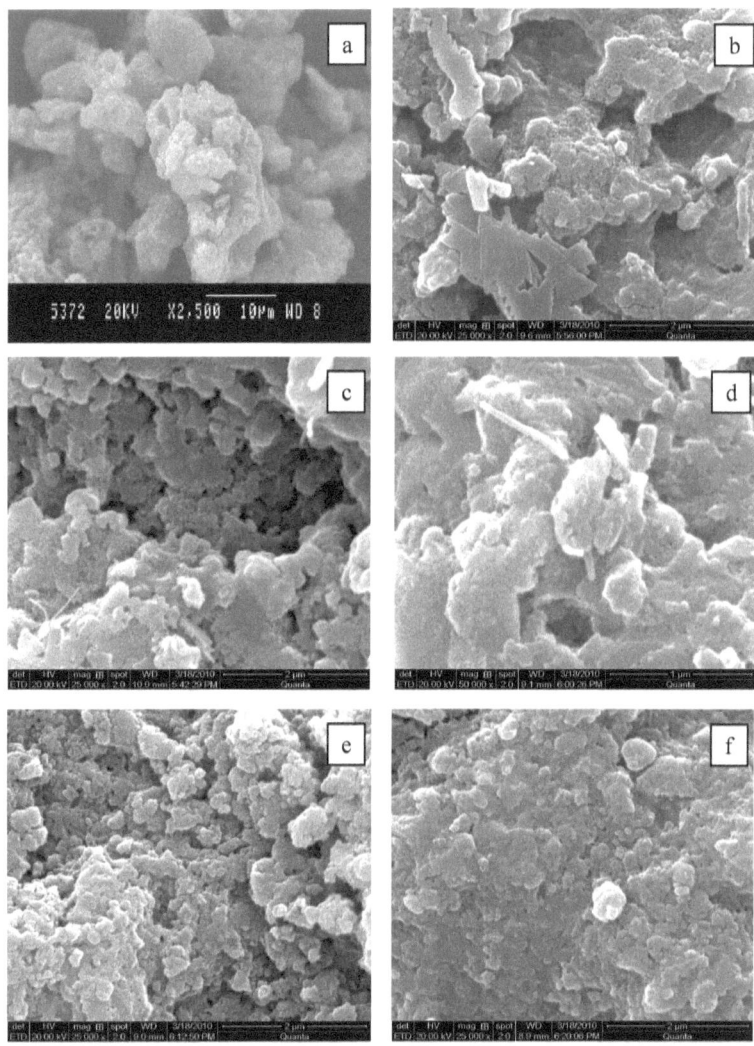

Figure3.3.2.(a-f) shows the SEM image of PANI and PANI/NiZnO₃ composites of different weight percentage (10, 20, 30, 40 and 50 wt%)

3.6.6. Polyaniline / CuO composites

The SEM image of 10 wt % of PANI/CuO composites shown in figure 3.3.3. (b) shows the grains are highly agglomerated, irregular in shape but they are well interconnected each others. The average grain size was found to be 0.22µm.

Figure3.3.3.(c) shows the SEM image of 20 wt % of PANI/CuO composites. The grains are irregular in structure, some of them are elongate and some are spherical in shape. The particles are individual and are separated with each other. The average grain size was found to be 0.37µm.

The SEM image of 30 wt % of PANI/CuO composites shown in figure3.3.3.(d) It is observed from the image that all grains are irregular in structure arranged one above the others. The average grain size was found to be 0.43µm.

Figure3.3.3.(e) shows the SEM image of 40 wt % of PANI/CuO composites prepared at room temperature. It is seen from the image the grains are clustered, have low porosity. The average grain size was found to be 0.47µm.

Figure3.3.3.(f) shows the SEM image of 50 wt % of PANI/CuO composites prepared at room temperature. It is found that from the image the grains are high agglomerated, have porosity and good interconnectivity between the particles. The average grain size was found to be 0.53µm.

By comparing the figure (a-f), it can be conclude that the gradual increase in granular size and change in morphology helps the transportation of charge particles through the carbon back-bone of polymer chains [18].

Figure3.3.3.(a-f) shows the SEM image of PANI and PANI/CuO composites of different weight percentage (10, 20, 30, 40 and 50 wt %)

Reference

1. B D Cullity, " Elements of X – ray diffraction" Addison Wesley pub. Co. Inc. Massachusetts (1956).

2. M J Burger, " X –ray Crystallography" Wiley, Newyork (1953).

3. A N Raransky, Ja, M. Struk, I. M. Fodchuk and N D Raranskay proc. SPIE – Int, Soc. Opt. Eng. (USA), 2108 (1993) 320.

4. B F Levine, Phy. Rev. B. 87 (1975) 2591.

5. H.Klug and Alexander, "X-ray Diffraction procedure" John Wiley NY(1974).

6. International Tanle of X-ray Crytsallogarphy,Pub., for inter union cryst. (The Kynouch press UK) 1(1952).

7. W L Bragg, The crystalline state, Macmillan, N.Y. (1993).

8. Jiang Xiaoming and Wu. Zigin. Chin. Phys. Lett. (China), 8 (1991) 356.

9. K C Nagpal, proc. Ind. Natl, Sci, Acd, A (India), 27 (1991) 325.

10. Narsimha Parvatikar, Syed Khasim, M. Revansiddappa, Shilpa Jain, S V Bhoraskar, and M V N Ambika Prasad Sens and Actuators B 114 (2006) 599.

11. Koppalkar.R. Anilkumar, Ameena Parveen, G.R. Badiger and M.V.N.Ambika Prasad., Physica B 404(2009)1664.

12. Alexander L. E, " X-ray Diffraction Methods in Polymer Science", John Wiley, New York, (1969)379.

13. K. Cheah, M. Forsyth, V. -T. Truong; Synth. Metals 101 (1999)19.

14. P. Lemon, J. Haigh; Mat. Research Bull. 34(1999) 665.

15. K. Cheah, M. Forsyth, V. -T. Truong; Synth. Metals 94 (1998) 215. Irkhin Yu P and Turov E A. Phys. JEPT, 33 (1957) 673.

16. Ramesh Patil, Aashis S. Roy, Koppalkar R. Anilkumar, Srikant Ekhelikar, J. App. Poly. Sci; (2011), DOI:10.1002/app.33600

17. Aashis. S. Roy, Machappa T, Sasikala.M.V.N and M. V. N. Ambika Prasad; Sensor letter, 9, (2011) 1-7

18. Aashis. S. Roy, Koppalkar.R.Anilkumar and   M.V.N.Ambika Prasad; J. Appl. Poly. Sci, 121(2011) 675

CHAPTURE IV

4.0. Introduction to Sensor Studies

Sometime earlier, the ultimate property of most polymers, which distinguishes them from metals, was their inability to conduct electricity. Electrical conductivity was introduced into these materials by adding conducting grains to form metal – polymer composite, or by the incorporation of ionic species to form polymer electrolytes. During the past 20 years a new group of organic polymers has been revealed with the ability to conduct electrical current inherently. These electroactive conducting plastics (ECPs) are still under development for appropriate applications, such as rechargeable batteries, capacitors, field effect transistors, EMI shielding, molecular magnets, enzymatic biosensors and more recently humidity and gas sensors [1]

Electroconducting conjugated polymers are good candidates for the elaboration of chemical or electrochemical sensors in two ways, either as a matrix for immobilization of active compounds or as sensitive compounds, where some properties film change in presence of the studied phenomena [2].

Conducting polymers can also be used to detect humidity and some gases and vapours by monitoring the change in conductance on exposure of the polymer [3]. Preliminary studies on these materials have shown that they exhibit a fast and reversible response even at room temperature.

A sensor is a device which detects or measures some condition or property and a record, indicated, or otherwise responds to the information received. Thus sensors have the function of converting a stimulus into a measured signal. The stimulus can be mechanical, thermal, electromagnetic, acoustic or chemical in origin, while the measured signal is typically electrical in nature, although pneumatic, hydraulic and optical signals may be employed.

There is an escalating need and desire for us to monitor all aspects of our environment in real time and this has been brought about by our increasing concerns with pollution, health and safety. There is also a desire to determine contaminants and analytes at lower and lower levels and one could say that the aim of all modern science is to lower the detection limits and to improve the accuracy and precision at those limits. Instrumentation has become so sophisticated that we are now able to detect chemicals in amounts smaller than we ever imagined of a few years ago.

Because of this desire and need of monitor everything around us there is a tremendous input of energy and resources into developing sensors for a multitude of applications. The end result of all this research will one day provide us with portable, miniature, and intelligent sensing devices to monitor almost anything we wish. For example, take our health; one can imagine in the future having a credit car size self-diagnostic unit with a multitude of chemical sensors and biosensors built into it so that we can monitor our well being at any instant.

In monitoring the environment one can imagine similar devices which could be used to test for say heavy metal pollutants in natural waters or the presence of bacteria in drinking water, swimming pools or at beaches. Bathers might carry such device with them to test the water before swimming. The possibilities are limitless and are controlled only by one's imaginations. The possibilities are really controlled by the Physics, Chemistry and Electronics of such devices and the art of that particular point in time and work on the principles of basic science [4].

Research activity especially on chemical sensors is now flourishing throughout the world [5]. Many papers of chemical sensors are being published in journals and read at domestic and international conferences. They convince us that chemical sensors are here to stay. Although various kinds of new devices and principles have been proposed not all of them have been commercially successful. Even scientifically fascinating and well-engineered devices sometimes find difficulty in the commercial market. Some of these encounter problems in the fact that reasonable productions are required for a successful device, not just high performance. Moreover, new devices must be introduced at the right time to meet social needs.

It was already known in the 1950's that metal oxides such as ZnO and NiO change semiconducting properties with change in partial pressure of oxygen, N_2O or other gases in the surrounding atmosphere. Relevant theories were proposed by many researchers understanding the nature of the gas-solid interactions as well as for controlling surface chemical processes such as catalysis. However, an approach in the opposite direction i.e. utilizing the phenomenon for the detection of gases was not conceived until 1962. In that year, Seiyama et al. from Japan reported that inflammable gases in air could be detected from a change in the electric resistance of a thin film on ZnO, while Taguchi claimed that a porous sintered block of SnO could also work in the same way. These findings clearly demonstrated the possibility of a sensing device based on an oxide semiconductor.

Despite such successful development in the past decades, however, fundamental understanding of the sensor remains far from being satisfactory. There is increasing need of new sensors capable of detecting humidity, various toxic gases and smell components. Trace gases sometimes at sub-ppm levels present in the environment or generated from food will be new targets of detection in the near future. It is unlikely that such demands will be met easily by simple extension of the present trial and error approaches. The introduction of a concept for the design of humidity and gas sensors is vital. Generally speaking, a sensor must posses at least two basic functions; i.e. a function to recognize humidity and a particular gas among others (receptor function) and another transducer the recognition into an electrical or optical signal (transducer function).

Classification of Sensors

Humidity Sensors: They are based on change in electrical properties of the material due to the absorption of water vapour. Hydrophilic polymers are used for resistance type humidity sensors, while hydrophobic polymers are preferred for capacitance type sensors.

(a)   Liquid and Solid electrolyte-electrochemical sensors: They are based on Faraday's law. Because of the ionic nature, the ionic conductivity in the electrolytes any current passing through it will carry a corresponding flux of matter. Therefore the measurement of pumping current provides an easy and accurate determination of the quantity of matter being transferred from one electrode to other.

(b)   Catalytic sensors: In catalytic sensor, the gases react on a catalytic filament via an exothermic process. The resulting temperature increase is being monitored by a corresponding resistance change in the filament.

(c)   Electronic conductive devices-semiconductor sensors: In which reversible reaction of the gas at the semiconductor surface results in a change of one of its electronic properties usually conductance.

4.1. Humidity Sensors

The importance of humidity sensing has been well understood and much research has been focused on the development of humidity sensitive materials. It has become evident in the recent years that the influence of humidity is of paramount importance in many types of areas; such as in moisture sensitive products, storage areas, computer rooms, hospitals, museums, libraries, high voltage engineering and accelerator systems. There has been a considerable interest in exploiting organic [6, 7] substances such as phthalocyanines and doped conducting polymers for humidity and gas sensing [8 - 11]. Advantages with polymers as sensing materials, is that they can be used at room temperature where the inorganic sensors normally operate at elevated temperature. The field of conducting polymers has been largely dominated by the search for higher conductivities, better stability, and greater processibility. The conducting polymers that have been widely investigated are polypyrrole, polythiophene and polyaniline. Polyaniline is unique among the class of electronically conducting polymers in that their electrical properties can reversibly be controlled by changing the oxidation state of main chain and by the protonation of amine nitrogen chain [12].

Various polymers have been used to prepare humidity sensors. From their basic principles, they are classified into two categories:

1. Based on the change in the electrical properties of the material due to the absorption of water vapours;

2. Based on the gravimetric change in the material such as the quartz crystal oscillator.

The first category is divided into two types, the electric resistance type and the capacitor type. Hydrophilic polymers are used for resistance type humidity sensors while; hydrophobic polymers are preferred for capacitance type sensors. Now a day, a variety of polymers doped with suitable molecules or dyes are being used for optical humidity sensing.

It is known that water increases the electrical conductivity of polyaniline through an increase in the interchain electron transfer [13 – 16] and / or by increasing the mobility of dopant ions [17]. Conductivity increases of several orders of magnitude in the presence of water have been recorded with linear and reversible responses.

Organic/inorganic composite materials have numerous bulk properties that can be improved and compared with those of the base polymers. In the present study we report humidity sensing and successful utilization of polyaniline – nickel zinc oxide, polyaniline – zinc ferrite composites and polyaniline – copper oxide as a humidity sensor.

The preparation of different samples with varying weight percent is explained in chapter II. Thus prepared $PANI/NiZnO_3$, $PANI/ ZnFe_2O_3$ and $PANI / CuO$ composites are used for humidity sensing.

4.2. Experimental Technique

Humidity Sensor chamber consists of side glass plates of 400 mm x 300 mm dimension and 6 mm thickness, provided with top & bottom plates made of fiberglass. The chamber is made airtight by rubber beading. This chamber houses sample holder and vaporizer. The vaporization of water will be controlled by varying the ac voltage in the dimmerstat, which is situated outside the chamber. The amount of water vapours present inside the chamber is measured with a hygrometer (Mextech-DT-615), which measures relative humidity (RH) along with temperature inside the chamber. A dc fan is provided, in order to distribute water vapours uniformly throughout the chamber, the electrometers used are digital meters of high accuracy. The block diagram of the sensor setup is shown in figure 4.1.The complete Photograph of the setup used for the measurement is shown in figure 4.2.

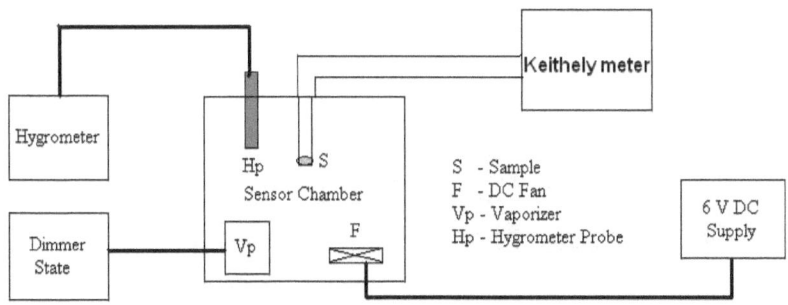

Figure 4.1 shows the block diagram of Sensor set up

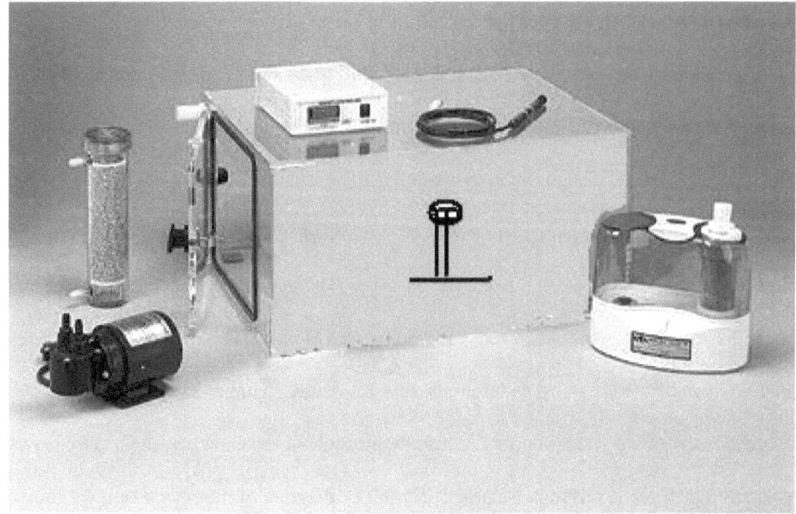

Figure 4.2 shows the Sensor set up used for the measurement of relative humidity

4.3. Results and Discussions

4.3.1. Polyaniline

Figure 4.3 shows the variation of sensitivity with change in relative humidity (%) for polyaniline. It is observed that at low humidity, the polymer is in the form of curl like a coil form, On the contrary, at high humidity, the polymer absorbs water molecules and the polymer chain swelling takes place, followed by the uncurling of the compact coil form into straight chains that are aligned with respect to each other which decrease the intergranular distance of two polyaniline islands. Hence, decrease of resistance with increase in the humidity proves the adsorption of water molecules, which makes the polymer more p – type in nature [18].

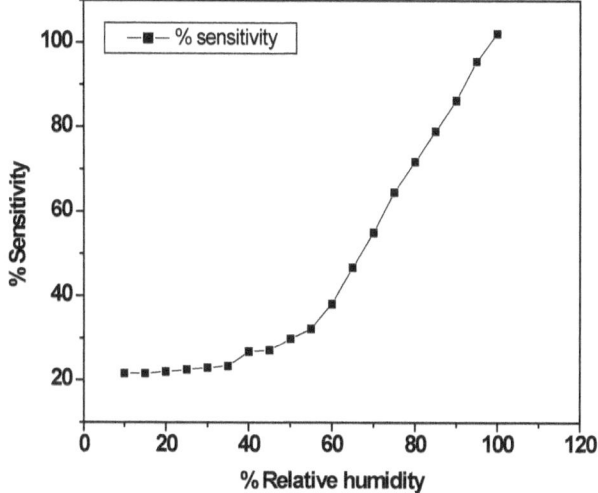

Figure 4.3 shows the variation of sensitivity with change in relative humidity (%) for polyaniline.

4.3.2. Polyaniline / ZnFe$_2$O$_4$ composite

Figure 4.4 shows the variations in sensitivity as a function of relative humidity (RH) value for PANI/ ZnFe$_2$O$_4$ composites in the form of pellets for three different wt% of zinc ferrite (10, 30 and 50 wt%) in polyaniline. The increase in the sensitivity with increasing humidity can be attributed to the capillary condensation of water vapors in all the pores. At low humidity, the capillary action is weak that restricted the flow of water vapors against the force of gravity. It is due to the weak intermolecular attractive force between the water molecule and composites surrounding surface. On the contrary, at high humidity, the polymer absorbs water molecules, and the polymer chains swelling takes place, followed by the uncurling of the compact coil form into straight chains that are aligned with respect to each other. On careful observation, it is clearly seen that PANI/ $ZnFe_2O_4$ composite shows a linear response from 10 to 95 % RH. Among all composites, 30 wt% of PANI / $ZnFe_2O_4$ composites shows high sensing response and exhibits good linearity in sensing response curve [19-22].

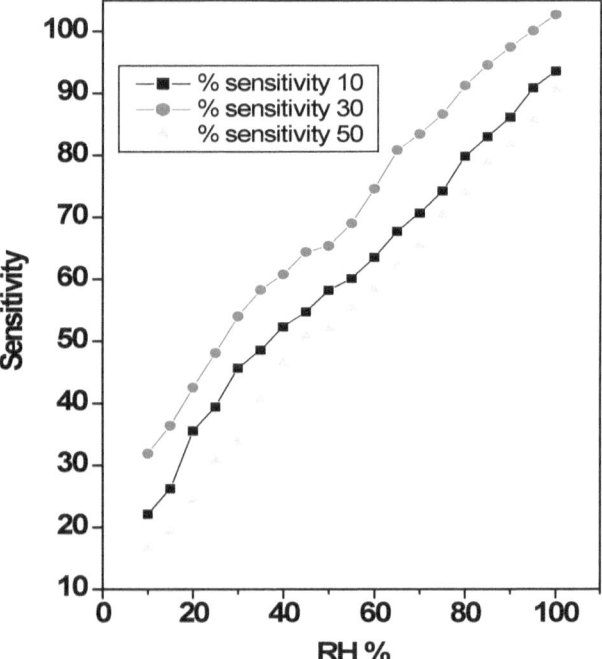

Figure 4.4 shows the variations in resistivity as a function of relative humidity (RH) value for PANI/ $ZnFe_2O_4$ composites of three different wt% of zinc ferrite (10, 30, and 50 wt %) in polyaniline.

4.3.3. Polyaniline / NiZnO$_3$ composite

Figure 4.5 shows the variations in sensitivity as a function of relative humidity (RH) value for PANI/ NiZnO$_3$ composites in the form of pellets for three different wt% of NiZnO$_3$ (10, 30 and 50 wt %) in polyaniline. The decrease in the resistance or increase in the sensitivity with increasing humidity can be attributed to the absorption of water molecule. At low humidity, the mobility of the NiZnO$_3$ ion is restricted because under dry conditions the polymer chains would tend to curl up into compact coil form. On the contrary, at high humidity, the polymer absorbs water molecules, and the polymer chains takes place, followed by the uncurling of the compact coil form into straight chains that are aligned with respect to each other. This geometry of the polymer is favorable for enhanced mobility of the NiZnO$_3$ ion or the charge transfer across the polymer chains and hence the conductivity. The almost linear variation with respect to percentage relative humidity (% RH) can be used in an amplifier circuit for converting the measured values into measurable % RH values. On careful observation of figure, it is clearly seen that PANI/ NiZnO$_3$ (10, 30 & 50 wt % of NiZnO$_3$ in PANI) composite shows an exponential response from 20 to 90 % RH. On the other hand, in PANI/ NiZnO$_3$ composite (30 and 50 wt % of NiZnO$_3$ in PANI) the sensitivity is found to increase from 40 up to 90 % RH in a exponential way. Thus PANI/ NiZnO$_3$ (10 wt % NiZnO$_3$ in PANI) shows better sensing properties and exhibits good linearity in sensing response curve [23-27].

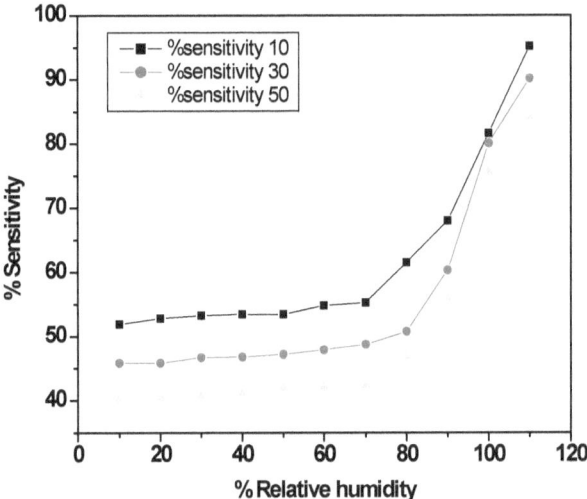

Figure 4.5 shows the variations in sensitivity as a function of relative humidity (RH) for PANI/ NiZnO$_3$ composites of three different wt% of nickel zinc oxide (10, 30, and 50 wt %) in polyaniline.

4.3.4. Polyaniline / CuO composites

Figure 4.6 shows that the change in the % sensitivity as a function of relative humidity of polyaniline / CuO composites for different 10, 30 and 50 weight percentages of CuO in polyaniline. It is observed that the % sensitivity of polyaniline / CuO composites increases with increase in relative humidity concentration in chamber. This is due to the absorption of water molecules causes swelling of the polymer chains, followed by the uncurling of the compact coil form into straight chains that are aligned with respect to each other. Among all composites, 30 wt % percentages shows the high sensitivity because it may be the critical concentration of MO in polyaniline matrix and porous morphology.

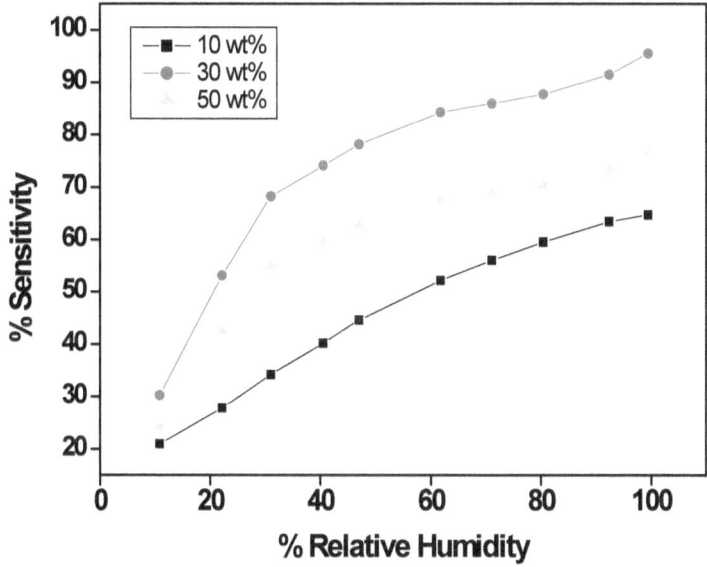

Figure 4.6 shows the variations in resistivity as a function of relative humidity (RH) value for PANI/ CuO composites of three different wt% of copper oxide (10, 30, and 50 wt %) in polyaniline.

Reference

1. Srivastava A, Rashmi and Jain K, Mater. Chem. & Phys. 105 (2007)385.

2. Wagh M S, Jain G H, Patil D R and Patil D R, Sens. Actuators B, 122 (2007)357.

3. Narsimha Parvatikar, Shilpa Jain, C M Kanamadi, B K Chougule, S V Bhoraskar, and M V N Ambika Prasad, Journal of Applied Polymer Science, 103 (2007)653.

4. W. Göpel, T.A. Jones, M. Kleitz, I. Lundström, and T. Seiyama (Eds),Chemical and biochemical Sensors, Part I, VCH, Weinheim, Germany, (1991)716.

5. T. Eklöv, Methods to improve the selectivity of gas sensors systems, Thesis at Linköping University, Sweden, (1999)188.

6. Prakash R Somani, R. Marimuthu, A.B. Mandale, Polymer, 42 (2001)2991.

7. Koenig J L, Spectroscopy of polymers (Amsterdam:Elsevier) 2nd ed (1999).

8. Pavia D L, Lampman G M and Kriz G S, Introduction to spectroscopy (Fort Worth: Harcourt College Publishers) 3rd ed (2001).

9. Suri, K.; Annapoorni, S.; Sarkar, A. K.; Tandon, R. P. Sens Actuators B, 81 (2002)277.

10. F.L. Lu, F. Wudl, M. Nowak, A.J. Heeger, J. Am. Chem. Soc. 108 (1986)8311.

11. Y. Sun, A.G. MacDiarmid, A.J. Epstein, J. Chem. Soc. Chem. Commun. (1990) 529.

12. Kobayashi, A.; Ishikawa, H.; Amano, K.; Satoh, M.; Hasegawa, E. J Appl Phys 74 (1993)296.

13. N. F. Mott. Philos. Mag. 19 (1969)835.

14. N. F. Mott and E. Davis, Clarendon Press, Oxford, 1979.

15. N. Yamazoe, Y. Shimizu, Sens. Actuators B, 10 (1986)379.

16. S.S. Pingale, S.F. Patil, M.P. Vinod, G. Pathak, K. Vijayamohanan, Mater. Chem. Phys. 46 (1996)72.

17. S. Hossein Hosseini1, S. Hossein Abdi Oskooei, and Ali Akbar Entezami, Iranian Polymer Journal, 14 (4), (2005)333.

18. Devidas Ramrao Patil and Lalchand Avachit Patil, IEEE sensors journal, 7 (2007)252.

19. P. Patil, J. M. Lee, Y.K. Seo, Y. K. Hwang, Y. K. Kwon, S. H. Jhung and J. S. Chang, J. Nanosci. Nanotechnol., 9 (2009)318.

20. M.L. Singla, Sajeela Awasthi, Alok Srivastava, Sensors and Actuators B: Chemical, 127, (2007) 580

21. Fuke MV, Kanitkar P, Kulkarni M, Kale BB, Aiyer RC. Talanta. 15 (2010)320

22. S Jain, Sanjay Chakane, A B Samui, V N Krishnamurthy, S V Bhoraskar, Sensors and Actuators B: Chemical 96, (2003) 124

23. Patil P, Lee JM, Seo YK, Hwang YK, Kwon YU, Jhung SH, Chang JS., J Nanosci Nanotechnol. 9 (2009) 318.

24. Peng Li, Yang Li, Lijie Hong, Yousi Chen, Mujie Yang, Materials Chemistry and Physics, 115, (2009) 395

25. Madhavi V Fuke, Prajakta Kanitkar, Milind Kulkarni, B B Kale, R C Aiyer, Talanta, 81 (2010) 320

26. Chih-Ting Lin; Che-Wei Huang; IEEE Sensors Journal, 10, (2010), 1142

27. Khan, Asif Ali; Khalid, Mohd.; Niwas, Ram, Science of Advanced Materials, 2, (2010) 474.

CHAPTER V

5.0. Summary and Conclusions

The advent of conducting polymers represents one of the important industrial revolutions of the current century. For more than a decade now, researchers have shown that certain class of polymers which are conjugated (those possess extended π – conjugation along their polymer backbone) exhibit conductivities varying from semi-conducting to metallic regime. An important fundamental property, i.e., electrical conductivity distinguishes polymers from metals. Polymers possessing high electrical conductivity are also referred as synthetic metals or conducting polymers offer a lot of advantages over the metals. These newly developed materials will not only replace metals in many areas, but also infiltrated our day-to-day life with a wide range of products extending from most common consumer goods like batteries to highly specialized applications in space and aeronautics.

Among conducting polymers, polyaniline family has attracted much attention of scientists world-wide because of their ease of synthesis, unique conduction mechanism, high environmental stability in the presence of oxygen and water, low cost, light weight and good sensing capability. They also exhibit highly reversible redox behavior, which is very important for many applications.

Extensive literature review suggests that the considerable efforts have been made by researchers all over the world in improving the conductivity of polyaniline by various doping techniques, but little is known about the dielectric properties of conducting polymers associated with conduction mechanism. It has also been noted through literature survey that the studies on dielectric and sensing properties of polyaniline composites is scarce.

88

To suggest any material to be used as a potential candidate for technological applications, it is essential for researchers to undertake basic studies governing electrical and sensing properties of such materials. Hence through this work author has made an attempt to tailor the electric and sensing properties of polyaniline by selecting appropriate materials as composites with polyaniline.

The important out come of the present work is summarized as follows:

i) In this present study, the author has synthesized successfully the conducting polymer, Polyaniline and its composites i.e. PANI / $ZnFe_2O_4$, PANI / $NiZnO_3$ and PANI / CuO in 10, 20, 30, 40 and 50 weight percentages.

ii) Through various characterization techniques employed on these composites, it is noticed that;

iii) X-ray diffraction pattern of polyaniline shows the presence of amorphous nature in polyaniline and a well ordered crystallinity in polyaniline composites.

iv) IR spectra confirms the homogeneous mixing of $ZnFe_2O_4$, $NiZnO_3$ and CuO in polyaniline, formation of new composites whose properties does not resembles either polyaniline or the individual materials used for preparation of composites. The characteristic stretching frequencies are shifted towards higher frequency side which indicates the homogeneous distribution of $ZnFe_2O_4$, $NiZnO_3$ and CuO particles in the polymeric chain. Further it confirms the Vander walls force of attraction between $ZnFe_2O_4$, $NiZnO_3$ and CuO with polymeric chain.

v) SEM micrographs show the presence of micro crystallinity in polyaniline and transformation from a branched pattern to highly granular structure (or a fibrillar morphology) of $ZnFe_2O_4$, $NiZnO_3$ and CuO in polyaniline.

vi) Studies are made on Humidity sensing properties and following conclusions are drawn.

vii) The almost exponential behavior of $ZnFe_2O_4$ and $NiZnO_3$ in polyaniline to the broad range of concentration of humidity proves to be promising materials as humidity sensors.

viii)   Because of the combination of electrical conduction of polyaniline with $ZnFe_2O_4$, $NiZnO_3$ and CuO these composites may find extensive technological applications which are to be explored in near future. Hence it is suggested through this work that

ix) Polyaniline – $ZnFe_2O_4$ and Polyaniline – $NiZnO_3$ composites are preferred for applications involving electrical properties and can be used as humidity sensors and high dielectric materials. Hence these composites may be used as microwave absorbing materials.

5.1. Possible Applications

a) On the basis of results of electrical and sensing properties, so obtained in polyaniline – $ZnFe_2O_4$ composites, they can be used as microwave absorbing materials and also in the fabrication of capacitors in high electric circuits.

b) Polyaniline – $NiZnO_3$ composites can be used as humidity sensors.

c) The high electrical conductivity of polyaniline and high dielectric constant CuO can be utilized to sense the volatile gases.

5. 2. Challenges in sensor research

a) The search and selection of proper materials, as well as improved and novel recognition mechanisms necessary for instant identification of a target component and the mechanism to create the signal will be obtained from the sensor.

b) The development of new materials for use as matrices to effectively immobilize receptor molecules to obtain stable and reproducible sensor function.

c) The development of solid-state versions of pH and ion-selective sensors.

d) Novel sensor substrates and internal electrodes for new planar fabrication designs to facilitate the use of advanced fabrication for automated sensor manufacturing, with the help of printing or semiconductor technologies for miniaturized sensor arrays, and improvements in signal processing technologies and instrumentation.

www.ingramcontent.com/pod-product-compliance
Lightning Source LLC
Chambersburg PA
CBHW030907180526
45163CB00004B/1744